BUILDING CONTRACT CONDITIONS

BUILDING CONTRACT CONDITIONS

ROBERT PORTER OBE, FCA, FCIS, IPFA

George Godwin Limited
The book publishing subsidiary
of The Builder Group

First published in 1980 by
George Godwin Limited
The book publishing subsidiary
of The Builder Group
1–3 Pemberton Row
Fleet Street
London EC4

British Library Cataloguing in Publication Data

Porter, Robert
Building contract conditions.
1. Building – Contracts and specifications –
Great Britain
I. Title
343′.41′078 KD1641

ISBN 0–7114–5509–0

While the principles discussed and the recommendations
given in this book are the product of careful study
and are believed to be accurate at the date of going
to press, neither the author nor the publisher can
accept any legal responsibility or liability for
any errors or omissions which may be made.

Typeset by Inforum Limited, Portsmouth
in 10 on 12 point Times New Roman
Printed and bound in Great Britain by
Tonbridge Printers, Tonbridge, Kent

CONTENTS

PREFACE

In the preface to my earlier book, *Guide to Building Contract Conditions*, published in 1975, I expressed the view that, since the building industry was currently faced with some 20 different forms of building contract, we might well have reached the stage where only the larger professional practices and contractors could afford the expertise to achieve some mastery of the subject.

In the intervening years some fundamental contract revisions have occurred and the forms have increased in number, length and complexity. In particular, 1980 sees the appearance of a number of major new forms.

The Joint Contracts Tribunal have now produced an extensively revised 1980 edition of the Standard Form, together with a completely new JCT nominated subcontract form which is to be used for all nominated work under the new main form. In addition, as a result of the growing importance of 'package deals', the JCT are also producing a new design and build form. The Minor Works Form has also been revised.

A new GC/Works/1 subcontract form is in preparation and the proposed contents of this have been outlined.

It is therefore a particularly appropriate time for a new book on building contracts, and the change of title and larger format of this current book give some indication of the extent of contract changes and the complexity of forms now available.

The book has been prepared to cover all building contract forms in general use, giving a clear explanation of the purpose and provisions of each form, and laying particular stress on points of difficulty.

It does not claim to be authoritative on the terms of the various contracts, nor is it intended to be a substitute for a careful study of these; but it should provide quick, accurate answers to the points which regularly arise on building contracts.

To facilitate this, numerous clause references are given in the text, but to avoid extensive cross-reference certain provisions have been repeated to make sections self-contained. This is supplemented by a detailed index.

For those wanting to compare new and old provisions of the various forms tables have been included giving subjects and clause numbers.

The contract forms dealt with are in general use throughout the United Kingdom, but in view of the differences in Scots law some amendments to meet these may be required. To illustrate this, in Chapter 3 there is an indication of the adaptations necessary when using the Standard Form of Building Contract in Scotland.

The developments over the past six years may add weight to the plea for a thorough review of contract procedures and provisions, if other than the larger organizations are to operate really effectively in the contracting field.

Robert Porter
Cambridge

ABBREVIATIONS

ACE	Association of Consulting Engineers
CASEC	Committee of Associations of Specialist Engineering Contractors
CITB	Construction Industry Training Board
DOE	Department of the Environment
EDC	Economic Development Committee (for Building)
EEC	European Economic Community
FASS	Federation of Associations of Specialists and Subcontractors
ICE	Institution of Civil Engineers
IOB	Institute of Building
JCT	Joint Contracts Tribunal
NEDO	National Economic Development Office
NFBTE	National Federation of Building Trades Employers
NJC	National Joint Council (for the Building Industry)
NJCC	National Joint Consultative Committee (of Architects, Quantity Surveyors and Builders)
NWR	National Working Rules (of the National Joint Council)
PC	Prime Cost (sum)
PSA	Property Services Agency (of the Department of the Environment)
RIBA	Royal Institute of British Architects
RICS	Royal Institution of Chartered Surveyors
SMM	Standard Method of Measurement (of Building Works)
VAT	Value Added Tax

1

SURVEY OF THE FORMS OF CONTRACT

1.1 Recent developments

In planning a book such as this, the seemingly endless stream of amendments and revisions to the twenty or more contract forms current in the building industry makes it difficult to choose a date when there is some degree of stability.

However, January 1980 is likely to be a milestone in the history of building contract conditions since in that month a new 1980 edition of the Standard Form of Building Contract was published which incorporates many recent amendments and substantial revisions (the previous edition was published in 1963, amendments having been issued annually over the following 16 years). In January 1980 also, the Joint Contracts Tribunal published for the first time its Nominated Subcontract Form which takes the place of the 'green' (NFBTE/FASS/CASEC) nominated subcontract form, for nominated subcontracts entered into under the 1980 Standard Form. The new forms will come into operation in mid 1980.

By that date several other major developments had taken place – the JCT Design and Build Form was awaited, the NFBTE subcontract form for use with GC/Works/1 was imminent, and the September 1977 edition of the Code of Procedure for Single Stage Selective Tendering was in operation. During the preceding three years or so, we had witnessed the introduction of Series 2 (48 category) Rules for Formula Adjustment (April 1977), (subsequently revised to align it to the 1980 forms of contract) the Construction Industry Tax Deduction Scheme (April 1977), subcontract set-off provisions (February 1976), emergency amendments to cover price code requirements (now withdrawn), a second edition of Government Contract Conditions GC/Works/1 (September 1977), and the Code of Procedures for Local Authority Housebuilding (March 1978). The sixth edition of the Standard Method of Measurement of Building Works was published in August 1978 and came into operation from 1 March 1979.

In addition to this there has been new legislation as well as some important case law on such matters as the responsibilities of architects, retention of title in goods and materials, limitation of actions, recovery of fluctuations etc.

This makes a formidable list and one may well ask if we have now reached the stage where the period of frequent and important changes has ceased and the end of the 'evolutionary process' reached. Unfortunately, a lengthy period of consolidation and education, although much needed, is unlikely.

Nevertheless, this is perhaps the best opportunity for years of setting down as lucidly, yet as briefly, as possible an understanding of the meaning and intention of current contract conditions in the building industry, although this is being done before there has been an opportunity to examine the new provisions in practice.

Fine shades of meaning and legal nuances are not considered in these pages but where contract provisions are related to each other this is indicated, where it would help in building up a better understanding of the subject. Where a point is the subject of argument and debate and has not been resolved in the courts, the view which finds most acceptance is set out. The effect of case law is incorporated in the text when appropriate without lengthy quotations or case references. The aim is to enable busy contractors, architects, surveyors, etc to check up on a point by quick reference to a clear, concise explanation. In striving for brevity one may fail to give a full evaluation of the issue but this risk must be taken if the book is to be both comprehensive and concise. However, the actual wording of each clause can be carefully studied from the detailed references and there are several learned and weighty textbooks on various aspects of the subject for those who have the time and the inclination to study any matter in depth.

It may be useful to consider at the outset why provisions are so complicated, sometimes obscure and subject to frequent alteration. There are at least five factors.

Firstly, the Joint Contracts Tribunal (JCT) which prepares and publishes the major forms most commonly used, has 11 constituent bodies (see Appendix B). To reconcile contrasting points of view and interests and secure the agreement of all parties calls for compromise and provisions more detailed and complex than those determined unilaterally (eg, GC/Works/1 Government Conditions of Contract – see Chapter 9).

Secondly, building organisation and techniques are never static and new developments need to be catered for (e.g, design and build projects, two stage tendering, price adjustment by formulae).

Thirdly, frequent and important decisions in the courts may call for amendments to conditions or prompt one or more of the constituent bodies of the JCT to initiate revisions to correct a situation, preserve a right, or safeguard an interest (eg, set-off, architects' responsibilities, etc).

Fourthly, legislation and Government policies increasingly impinge on building operations and may require incorporation in or amendment to existing conditions (eg, VAT, the construction industry tax deduction scheme, etc).

Fifthly, the members of the Joint Contracts Tribunal, steeped in contract matters, are constantly striving to perfect contract conditions so that the standard form, in particular, is really constantly under review. The danger is, of course, that a 'perfect' set of conditions may fail to meet the essential requirement of the users for whom they are intended – lucidity and comprehensibility.

1.2 Contracting arrangements

In building contract procedures it is dangerous to generalise but it can be said that building contract conditions are based on the principle that the employer (or building owner) enters into a contract with the contractor who is to carry out defined work for an agreed price. The contract documents normally include drawings, bills of quantities or, failing that, a specification (and perhaps a schedule of rates), as well as the contract conditions themselves. Under the Standard Form of Building Contract the architect who is appointed to design and supervise the work is named in the articles of agreement, but he is not a party to the contract although his powers and duties are very extensive. Where bills of quantities are provided, the quantity surveyor is named in the articles of agreement, and his duties are described in the contract conditions. If a consultant (eg, an engineer) is employed, this is a matter between the employer and his architect.

Work, however, may be carried out without the services of an architect and suitable conditions of contract have been prepared to meet a variety of circumstances such as private housebuilding, repairs and maintenance.

The range of the contracts covered is described broadly in this introduction and detailed consideration of each form of contract is dealt with in the following chapters. The Appendices at the end of the book provide a wide range of additional information.

Tendering practice and procedures are dealt with first in Chapter 2 since these are closely linked with contract conditions on which tenders are based. Some new concepts and procedures have been introduced into tendering in the past five years and this subject merits careful study. The contract documents at the tendering stage set out the conditions for carrying out the work and determine the future contractual relationships between the parties. The forms of contract published and current in January 1980 are then reviewed and explained as follows.

1.3 Standard Form of Building Contract *(see Chapter 3)*

This form, published by the JCT, is by far the most important for work carried out under the supervision of an architect, other than Government contracts. It is widely used and its provisions should be fully known and appreciated by architect, surveyor and builder alike. To avoid an unduly lengthy and complicated chapter two aspects of the contract are chosen for special treatment. Chapter 4 is devoted to firm and variable price contract provisions, including formula adjustment, and Chapters 5 and 6 deal with the nominated subcontract conditions – old and new style. The date for the operation of the new edition is mid 1980.

1.4 Firm price and variable price contracts *(see Chapter 4)*

The provisions for fluctuations have become more complicated not only because of the widening of the scope of Standard Form Clauses 37-40 (formerly 31) and related clauses, but by the introduction in April 1977 of Series 2 Price Adjustment Formula with 48 categories, compared with the 34 in Series 1. Price code requirements no longer govern recovery provisions under the Standard Form, but in view of their significance there is a historical note on the subject.

1.5 Subcontract conditions – NFBTE/FASS/CASEC forms *(see Chapter 5)*

Under the Standard Form of Building Contract, subcontractors fall into two categories and a suitable form of subcontract is available for each. This chapter is concerned with the nominated subcontract under the pre-1980 Standard Form (the 'green' NFBTE/FASS/CASEC form) and the continuing arrangements for non-nominated (now known as 'domestic') subcontractors under the 'blue' form. (Because of major changes in the 1980 Standard Form affecting domestic subcontractors, certain significant amendments to the 'blue' form will be necessary.)

(i) For subcontractors nominated by the architect under clause 27 of the pre 1980 Standard Form of Building Contract, the Standard Form of Subcontract (NFBTE/FASS/CASEC 'green' form) was widely used, as it incorporated the conditions in clause 27 under which the nomination took place. Although in nominated subcontracts let under the 1980 edition of the Standard Form the JCT Nominated Subcontract Forms must be used, nominated subcontracts entered into under the pre-1980 Standard Form, and carried out under the 'green form', will still have a considerable period to run although the form will be 'phased out'. For this reason both the 'green' form and the JCT subcontract form come within the ambit of this guide. Chapter 6 is devoted to the new JCT Nominated Subcontract Forms which come into operation in mid 1980.

(ii) For (domestic) subcontractors to whom work is sublet by the main contractor under clause 19 (formerly 17) of the Standard Form of Building Contract, the non-nominated subcontract ('blue') form, giving suitable protection to the parties, should still be used. The JCT has not concerned itself with these conditions so the 'blue' form, with appropriate revision will continue under the 1980 edition of the Standard Form as it did under the earlier edition.

In both instances under the appropriate subcontract conditions the parties are the main contractor and the subcontractor and the employer has no contractual relationship with subcontractors. In the past such a relationship could have been created in the case of nominated subcontractors by the use of the RIBA Form of Agreement which, since it could apply to contracts still current, is considered in Chapter 5. From January 1980 Agreement NSC/2 (or NSC/2a), part of the new JCT Nominated Subcontract documents, is available with generally similar effect (see section 6.2.1).

1.6 Nominated subcontract conditions – the JCT form *(see Chapter 6)*

These new documents were published in January 1980 and are to be used for nominated subcontracts entered into under main contracts let on the 1980 edition of the Standard Form. They consist of the following forms detailed in clause 35.3 of the Standard Form – Tender NSC/1; Agreement NSC/2 or 2a; Nomination NSC/3; and Subcontract NSC/4 or 4a. With the introduction of this first JCT form to deal with subcontract conditions, the Standard Form of Building Contract was substantially altered in the 1980 edition, clause 35 of the main form now containing the nominated subcontract provisions, to integrate with those of JCT subcontract forms and procedures. This clause (its counterpart being clause 27 of the pre-1980 edition) is examined, along with the subcontract provisions themselves and the extensive attendant procedures. An opportunity is taken of comparing these new provisions with those of the earlier 'green' form (see Chapter 5 above) because it is of value to relate new provisions to earlier ones which generally are well known and understood.

1.7 Other Joint Contracts Tribunal forms *(see Chapter 7)*

There are five JCT forms, based on the Standard Form of Building Contract and adapted to meet special circumstances.

(i) *Fixed Fee Form*
The contract sum is based on prime cost as defined, plus a fixed fee to cover overheads and profit.

(ii) *Minor Works Form*
This is a simplified version of the Standard Form used for small jobs on a lump sum basis where bills of quantities are not provided, and revised in 1980.

(iii) *Renovation Grant Work*
There are two forms available – one where an architect is employed, and one where no architect is employed.

(iv) *Nominated Suppliers' Tender Form*
Although not mandatory, this new JCT form could be used with considerable advantage in the nomination of suppliers under clause 36 of the 1980 edition of the Standard Form.

(v) *Design and Build Forms*
For the first time, the JCT published in 1980 conditions for design and build contracts as follows:
(a) Standard Form of Contract with Contractor's Design ('Design and Build Form')
(b) Clauses modifying the Standard Form for use with Quantities and Contractor's Proposals ('Contractor's Designed Portion').

1.8 Miscellaneous forms and publications *(see Chapter 8)*

A variety of forms have been prepared by various organisations as detailed in the text to cater for particular needs.

(i) *NFBTE Design and Build Form*
This agreement between employer and builder was published by the National Federation of Building Trades Employers (NFBTE) in 1970 to cover a 'package deal' where the builder was responsible for design. It is superseded by the 1980 JCT Design and Build Form but it may be applicable to some existing contracts.

(ii) *Labour only subcontract form*
This concerns work sublet by the main contractor under clause 19, or otherwise on this basis and is published by the NFBTE.

(iii) *Nominated suppliers*
There is a form of tender for suppliers nominated by the architect under clause 28 of the pre-1980 Standard Form. The RIBA Warranty Form is also reviewed. The JCT Nominated Suppliers' Tender Form under the 1980 Standard Form is dealt with in Chapter 7.

(iv) *New work (including housebuilding)*
This is used where no architect is employed.

(v) *Construction of new streets*
This is a special NFBTE form of agreement between local authority and housebuilder under the Highways Act 1959.

(vi) *Conditions of estimate*
Used for smaller jobbing and maintenance work, these simple conditions prepared by the NFBTE can avoid trouble and dispute.

(vii) *Daywork – definitions, etc*
Although this is an adjunct to contract conditions (eg, under clause 13 of the Standard Form) it is important enough to be included under this heading. (published jointly by RICS and NFBTE)

(viii) *GC/Works/1 Subcontract Form*
This new form being prepared by the NFBTE can be used for both nominated and non-nominated sub-contractors under GC/Works/1.

1.9 Government building contract conditions *(see Chapter 9)*

The conditions in general use are found in Form GC/Works/1 which superseded CCC/Works/1 in March 1974. The second edition of GC/Works/1, 'General Conditions of Government Contracts for Building and Civil Engineering Works', was published in September 1977 and its provisions and the alterations since the first edition are fully reviewed.

Since the GC/Works/1 conditions differ substantially from those in the Standard Form, some of the salient differences are highlighted. In the absence of a Government subcontract form, the NFBTE is publishing a suitable form for use in connection with GC/Works/1 and this is referred to in Chapter 8. The special supplementary conditions on incomes policy introduced in March 1978 are only briefly referred to since they were withdrawn in December 1978 (see section 4.3.2).

There is also a brief review of GC/Works/2, used generally for work under £30,000 on specification and drawings only. Measured Term Contracts (Form C1501) for maintenance and other work in Government establishments call for special provisions which are considered in some detail because although this work is carried out by relatively few contractors who specialise in this field, the value of work executed is substantial.

It is hoped that this approach will present a logical pattern for what is a very complicated network of forms and procedures. Additional information on related matters is to be found in the Appendices, and the index has been made as comprehensive as possible to facilitate quick reference to any significant point.

2

TENDERING PROCEDURES

Tendering for building work requires a sound knowledge of procedure and practice and an appreciation of some basic considerations related to the law of contract. This chapter is concerned primarily with tenders for new work and other sizeable projects under the Standard Form of Building Contract. For minor repair and maintenance work, domestic decorations and so on, a much more simple procedure should be adopted although certain basic principles would still apply. Building contract conditions and procedures must be considered against the legal background which can be studied in authoritative works on the law of contract etc. These wider issues do not come within the scope of this guide.

2.1 The pre-tendering stage

When building work of any reasonable size is required, an architect is usually commissioned. He analyses his client's requirements, prepares an appraisal of the project, including planning considerations, as a basis for the client's decision, instructs quantity surveyors and consultants[1] where necessary, prepares outline proposals, timetables and cost estimates. Once all the necessary decisions have been taken and planning approvals, etc obtained, drawings, bills of quantities and perhaps a specification are prepared to enable tenders to be invited for the work.

Tenders can be invited in several ways:

(a) By advertisement briefly describing the project and inviting those interested to apply for tender documents. If tendering is open to all suitable firms who apply, this could mean 20 or more tenderers with commensurately higher cost to the industry and, in due course, its customers.

(b) By direct selection or by advertisement stating that a selection will be made from firms replying and considered suitable (so that four or six firms might be selected, according to the size of the contract).

[1] Scale of fees for professional services are published by:
The Royal Institute of British Architects, 66 Portland Place, London WI.
The Royal Institution of Chartered Surveyors, 12 Great George Street, London SWI.
The Association of Consulting Engineers, 87 Vincent Square, London SWI.

(c) By inviting firms on a standing list. This is the practice of some large public authorities, whose lists may be subdivided into contract sizes and areas of operation. Lists are reviewed regularly, usually after advertisement. Firms are promoted, and added to or removed from the lists. If there are more firms in any category than would normally be invited, they are selected in rotation often after a preliminary enquiry so that only those wishing to tender are included. Fairness throughout is the keynote.

(d) By selecting a firm at the outset without competition, because of its expertise, standing, previous satisfactory work or for some other reason. Where this happens the association of the contractor with the professional team and the building owner at the earliest stages can make for excellent team work, provided that knowledge and experience are freely pooled.

(e) By serial tendering to ensure continuity of work with its attendant advantages. This usually involves the letting of one contract by competition, with others to follow, based on original prices suitably updated or renegotiated.

(f) By two stage tendering, generally for large and complex projects where it is of special benefit to bring in the contractor at an early stage (possible procedures are reviewed later in this chapter).

There is one rule which, if observed, can ensure more effective tendering and prevent ill feeling. If invited to tender but unable to do so, the contractor should inform the architect at once. If, after receiving the tendering documents, it is no longer possible to tender, the documents should be returned immediately with an explanation. If tendering is complex and the time is considered inadequate, an extension of tendering time could be granted to all tenderers on application.

2.2 Submitting tenders

There are frequent misunderstandings about the legal considerations affecting tenders. A tendering invitation is an invitation to offer based on the tendering documents.

The tender is an offer. If it is not strictly in accordance with the tendering documents it becomes a qualified offer which can be rejected. For example, a tender invited with a 15 month contract period might on submission be qualified by stipulating an 18 month period. It could quite properly be ruled out when tenders were being considered. If the contractor is not entirely happy with some points, a note accompanying his tender suggesting discussion of them should his tender be considered would be politic at this stage (see below (2.3.1) for procedures for qualified tenders). However, it cannot be emphasized too often that the tendering documents form the basis for the offer. Once accepted they virtually determine the conditions for carrying out the contract. They should be fully examined at once, a careful watch being kept on the following points, which, although relating to the Standard Form, are of general application.

(i) alterations to the Standard Form or the introduction of special conditions.
(ii) apparent discrepancies between drawings and bills
(iii) departure in bills from SMM requirements, if applicable
(iv) points calling for clarification
(v) 'interpretations' of the conditions of contract by one party which may be wrong
(vi) the provisions to be included in the Appendix to the Standard Form eg, damages, dates for possession and completion, defects liability period, retention, period for honouring interim certificates, extent of public liability cover, fluctuations, formula rules, etc
(vii) treatment of optional clauses
(viii) any special insurance requirements
(ix) any endeavour to widen contractors' responsibilities in relation to design, nominated subcontractors, etc.

It is vital to satisfactory tendering and to establishing the maximum confidence between the parties that the standard conditions of contract which are widely used and understood, should be used without amendment.

It is important, too, that all points of doubt be cleared up before the tender is submitted. It is ill-advised to tender where a condition or requirement is provisional, obscure, unacceptable or undesirable, and to hope that if the tender is successful, the point can be dealt with later and the terms of the contract changed.

A tender submitted with a time limit (eg, 'if the tender is not accepted by (date) this offer lapses') should not be considered a qualification where it is not at variance with any provision in the tender documents. It should, however, safeguard a contractor at an unstable time where circumstances are changing rapidly. All too often tenders are submitted and overlooked if acceptances are delayed. Positive action is needed since only if there is an abnormal delay in acceptance can a tender be claimed as having automatically lapsed. Once the tender is accepted without qualification, a contract is made, and withdrawal thereafter could constitute a breach of contract with a claim for damages resulting (eg, the difference between the accepted tender and the next lowest, or a subsequent tender necessarily sought). It is a common fallacy that even when a tender is accepted the contract is not made until the contract agreement is actually signed. This is the case only if it is made a condition at the outset. As a corollary, even if through administrative negligence, as often happens, an agreement is not eventually signed, the parties are generally bound by its terms if the tender is submitted under these conditions. An offer can be withdrawn after a qualified acceptance (eg, 'subject to Ministry approval') and before final (unqualified) acceptance.

Contractors are sometimes required to undertake to keep the tender open for acceptance for a stipulated period. This is not binding, and tenders can be withdrawn at any time before acceptance.

As in all tendering matters, agreed procedures should be punctiliously observed; the opening of tenders should follow the precise pattern described later.

2.3 Selective tendering

If all suitable firms wishing to tender are allowed to do so, too many might apply. Tendering is expensive and the overall cost to the industry and its customers has to be borne in mind. At the same time if firms are not allowed to tender there might be a feeling that something underhand is going on.

Is it possible to select firms for work on fair and clearly agreed rules to meet both of these points? Can this selection also prevent unsuitable firms from appearing on tender lists?

These points were considered by the Howard Robertson Committee on Tendering Procedure in 1954 which recommended the universal adoption of selective tendering and set out principles for the necessary procedures.

2.3.1 SINGLE STAGE SELECTIVE TENDERING

In 1959 the NJCC produced a Code of Procedure for Selective Tendering which was revised in 1969 in the light of the 1964 Banwell Report recommendations, and revised yet again in 1972 in collaboration with the Department of the Environment. A new edition was published in 1977, taking into account the 1975 NEDO report on 'The Public Client and the Construction Industry'. It includes some significant amendments, including criteria for selection and procedures for dealing with qualified tenders. The code covers the United Kingdom and although not mandatory, is widely observed for public authority and private work, and is supported by the DOE. Although it assumes the use of the Standard Form of Building Contract unamended, and indeed strongly recommends this, it is relevant where other forms of contract are contemplated.

The 1977 edition is entitled 'Code of Procedure for Single Stage Selective Tendering' in view of the NJCC decision to publish in due course a new 'Two Stage Tendering Code' (see below). Reference is made in the Code to the EEC directive on tendering which is set out later in this chapter (2.5).

The main provisions of the code are as follows:

(i) The maximum number of tenderers is related to contract size the following scale to be a guide:

Up to £50,000 in value	5 tenderers
£50,000 to £250,000	6 tenderers
£250,000 to £1 million	8 tenderers
£1 million +	6 tenderers

Firms selected either from an approved list or an ad hoc list should all be capable of doing the job by reference to financial standing, management structure, capacity, experience and reputation. A preliminary enquiry with all relevant information, as set out in Appendix A of the code, ensures a willingness to tender, firms not selected to be promptly informed. A reasonable time (eg, four to six weeks) should be allowed before the despatch of the tender documents. A contractor wishing to withdraw should notify the architect at once. Only in very exceptional circumstances should a tender other than the lowest be accepted. This is always subject to the understanding that the building owner is not obliged to accept the lowest or any tender.

(ii) The conditions of tendering should be made absolutely clear. The contract period should be specified in the tender documents and not made the subject of competition. Offers on alternative periods of completion should only be on invitation to all, at the time of tendering, and these should all be submitted at the same time.

(iii) Alterations to the Standard Form should be avoided but if these are absolutely necessary full details, with reasons, must be set out in the preliminary invitation to tender.

(iv) Four working weeks at least should be allowed for tendering – longer for complicated or larger projects. The date and time for submission should be given and late tenders returned unopened. The prompt opening of tenders, and the notification of tender prices is stressed. The lowest tenderer should be asked to submit his priced bills within four days. A tender may be withdrawn at any time before acceptance.

(v) Prompt notification of results should be by a list of tender prices. To safeguard the building owner's interests in the event of withdrawal, the second and third lowest tenderers should be so informed and told that they will be approached subsequently if their offers are to be further considered. A new provision emphasises the need for tenders to be based on identical tender documents with no attempt to qualify tenders. The architect should be informed at once of conditions regarded as unacceptable, or of points requiring clarification, etc. The architect should inform all tenderers of any resultant amendment and extend the time for tendering. Otherwise a tenderer should be given the opportunity to withdraw any qualification. If he fails to do so the tender should be rejected if the qualification is considered to give him an unfair advantage.

(vi) If, after examination of the contractor's bills by the quantity surveyor, an error is discovered, it should be reported to the architect who, with the building owner, will determine the course of action as previously decided and indicated in the tender documents. There are two alternatives, the one chosen to be stated at the tendering stage. In the first the contractor should be asked to confirm his offer with the bill rates adjusted pro-rata to the error, an endorsement signed by both parties to be added to the priced bills. If he withdraws completely, the second lowest tender would then be considered.

In the second alternative the contractor should be asked to confirm his offer or amend it for genuine errors. If it is amended and is no longer the lowest, the next lowest offer would be considered. Local authorities have been advised by the DOE that this alternative is not 'suitable' for them (see 2.4.4 below).

(vii) If a tender is above the budgeted figure, a revised price should be negotiated with the lowest tenderer, the basis and any agreement reached to be fully documented. If these negotiations are unsuccessful, then they should take place with the second lowest and then with the third lowest tenderer. If all negotiations fail, new tenders may be called for.

The code refers to the importance of properly judged starting dates.

The NJCC has published as an appendix to this code, the 'Local Authority Tendering Questionnaire', suitably adapted for use in the private sector.

2.3.2 TWO STAGE TENDERING

NJCC Code of procedure

The National Joint Consultative Committee is preparing a Code of Procedure for Two Stage Tendering which may be published in 1980.

This may deal with tendering procedure only and not with a contractor's possible design involvement. In large and complex schemes close collaboration between contractor and architect at the design stage could be advantageous and if the contractor were involved in the planning of the project, benefit should result.

Under two stage procedure a contractor is selected at the first stage on documents sufficient to provide a competitive basis for selection, establishing at the same time the broad principles of layout and design as a basis for detailed development later without duplication of work.

The procedures for the first stage would be similar to those in the single stage code referred to above, on such matters as selection of tenderers, preliminary enquiries, tender documents including an explanation of two stage procedure, time for tendering, qualification, errors, evaluation of tenders, acceptance and notification of results.

It is likely that approximate bills would be used for stage one, and in evaluing tenders it would be necessary by adjustments in tender sums to bring all tenders on to a comparable basis. It should then be possible to select the contractor for the second stage negotiations but no contract would be made until these were completed.

The second stage would be mainly concerned with developing the project design, preparing bills of quantities to be priced on the basis of the first stage tender, and agreeing a final price for the work when the contract would then be made.

If a project were cancelled before the second stage was completed, the contractor could reasonably expect to recover any costs resulting from any contribution he had made to the design. He would have to bear, however, tendering costs for stage one.

The recommended use of the Standard Form of Building Contract without amendment is likely, although as in the single stage code the procedures could apply whichever contract conditions were used.

Local authority procedures
The Department of the Environment has already given guidance on two stage tendering to local authorities in Chapter 6 (Annex 6B) (May 1977) of its Code of Procedures for Local Authority Housebuilding, dealt with in the following section. The main points on two stage tendering procedures are set out below. This is primarily concerned with housing but it lays down general principles which may not differ much from those in the promised NJCC code.

General arrangements: The purpose of the first stage is to select a contractor and provide a level of pricing for subsequent negotiation. In stage two, in collaboration with the selected contractor, drawings and specification are finalised enabling comprehensive bills to be prepared for pricing and subsequent agreement of the contract price.

Procedure: The tender documents at the first stage would give information on housing layout and types and enough details to provide a basis for pricing. In the invitation to tender to selected contractors there would be an explanation of the two stage procedure, details of the project and programme, the method of tendering and the closing date. The tender documents would consist of layout and sketch plan with sections and elevations, the specification and, desirably, approximate or notional bills. A preliminaries bill as for single stage tendering should be supplied. Contract conditions should be the standard form unamended. The object is to provide a sound basis for selection and for the agreement of the contract sum at the second stage.

In the second stage working drawings should be prepared, and bills finalised for pricing jointly by the contractor and the authority's quantity surveyor on the basis of first stage prices. If this is not followed up promptly, some allowance may have to be made for inflation. Although the contractor may contribute to design matters as well as construction organisation, the design responsibility should rest entirely on the local authority. Annex 6C does in fact set out procedures for two stage design and build tendering.

2.4 Code of procedures for local authority housebuilding

In September 1976 the Department of the Environment announced that this code was being prepared for everyday use by members of councils and their officers. Its purpose according to the 'general introduction' (Chapter 1) is to assist local housing authorities in obtaining value for money in their housing programmes and 'maintaining adequate safeguards against malpractice'. The code consolidates and brings up-to-date advice given in earlier Government circulars and in other publications (eg, NJCC Single Stage Selective Tendering Code). The code does not cover such matters as financing, subsidies, planning, etc.

The code is divided into the following chapters:

1. General Introduction
2. The Client
3. Client's Role in a Single Project
4. Client's Role in the Housing Programme
5. Client's Role in Review of Housebuilding Policies
6. Procurement Procedures
7. Department Cost Control in Design in Individual Schemes
8. Main Contractor – Selection and Appointment
9. Firm Price and Fluctuation Contracts
10. Contract Conditions
11. Execution of Contract

The first part of the code which appeared in September 1976 covered Chapter 1, 'General Introduction', and Chapters 8 to 10 on tendering and contract conditions. Chapter 6 on 'Procurement Procedures' was issued in May 1977 and in March 1978 revisions of earlier chapters and the publication of Chapters 2,3,4,5,7 and 11 completed the code which superseded a substantial number of departmental circulars listed in DOE Circular 22/78 on 17 March, 1978. The provisions at October 1979 which are particularly relevant to the subject matter of this book can be summarised as follows:

2.4.1 ROLE OF THE CLIENT *(CHAPTERS 2 – 5)*

These chapters, produced with the assistance of an advisory group of local authority chief officers, are designed to help officers and members of councils in fulfilling the role of client in the housebuilding process. The subject matter is primarily concerned with local authority organisation and since this is not immediately relevant to construction contracts, is only referred to briefly here.

Chapter 2 refers to post war housing problems and the continuous efforts needed for a solution. The importance of development teams with clearly defined functions is underlined.

The client's role is analysed in Chapters 3, 4 and 5 under the headings of decision making and management tasks. Annex 3A sets out for single projects a description of procedures and information required by the client on preliminary work, design, procurement (single stage and two-stage procedures) and construction. Annex 3B is concerned with the composition of the project development team and the selection of consultants, and 3C with the evolution and contents of the design brief.

Chapter 4, dealing with the client's role in the housing programme (long and short term), lays emphasis on the work of the programme development team. Annex 4A deals with procedures, initial definition of targets, their review, adjustment and final definition, and the procedure for adjustment during construction. The formulation and monitoring of housebuilding policies are dealt with in Chapter 5. The necessary decisions and management tasks are examined in relation to the composition and function of policy development and monitoring teams. Annex 5A details tasks to be undertaken, information required and procedures to be followed in response to a changing situation, to the appraisals of programmes and projects, and to the results of monitoring reviews.

2.4.2 PROCUREMENT PROCEDURES *(CHAPTER 6)*

The procedures for placing contracts are reviewed, decisions on the method employed to reflect the size and complexity of the project, type of building and method of construction, considerations of time, professional resources and pre-contract involvement:

(a) *Traditional single stage tendering*
where planning and design responsibility rests fully with the client's professional team – suitable for all sizes of projects (fully detailed in Annex 6A)

(b) *Traditional two-stage tendering*
where a contractor is selected before large and complex schemes are fully designed so that use can be made of his expertise (fully detailed in Annex 6B)

(c) *Design and Build tendering*
where responsibility for design is shared between the client's professional team and the contractor (fully detailed in Annex 6C)

(d) *Package deal*
where full responsibility for design documentation and construction rests on the builder.

It is emphasised that in (a) and (b) the appropriate standard form should be used without modification. In Annex 6A there is reference to the Code of Procedure for Single Stage Selective Tendering 1977 (see 2.3.1 above).

There is guidance on continuation contracts where a subsequent contract is let on the basis of the prices for a previous contract obtained in competition. There is also a reference to the grouping of schemes and to serial contracts let on the basis of prices contained in a standing offer for a guaranteed amount of work of a similar nature in a limited period.

2.4.3 COST CONTROL *(CHAPTER 7)*

This chapter is concerned with cost planning and the housing cost yardstick.

To ensure that public funds are put to the best use, cost planning is vital. Cost targets should be set for the design of each project built up from cost data.

From the initial consideration of alternative schemes, through all the design stages, cost planning and control must be exercised so as to ensure that tenders are within the expenditure limit.

There is a useful section on control through the housing cost yardstick which controls capital expenditure, limits the subsidy and provides a cost planning tool.

Based on a wide-ranging study of housing costs, including the state of the market and the current level of tenders, the yardstick (which excludes site development works) represents the admissible cost limit for subsidy. Up to 10 per cent above the yardstick may be incurred but there is no subsidy for this excess. Adjustments are made for regional variations, accommodation for old, single, disabled, etc, play space, small schemes and above average car accommodation. A higher yardstick may be specially approved for noise factors and for materials, special foundations, or design. A maximum of 5 per cent above yardstick is allowed for redevelopment sites.

2.4.4 MAIN CONTRACTOR SELECTION AND APPOINTMENT *(CHAPTER 8)*

Standing orders are required by law to provide for competition and to regulate invitations to tender. Tendering procedure should ensure competition, fair awards and value for money. Of the three recognised procedures, open competition, selective tendering and negotiation based on a contract won in competition, open tendering is not recommended. It is open to the following possible objections – low prices and bad building, uneconomic prices leading to failure with added costs and delays, high cost of overall tendering and increased authority staff costs; unsuitable tenderers may have to be passed over.

Selective tendering

Selective tendering is advocated, the NJCC Code of Procedure for Single Stage Selective Tendering 1977 to be followed subject to certain amplification (*Note*: the references below are to clauses in the NJCC code): eg, recommended number of tenderers should be the maximum (c3); all eligible firms should be given the chance to tender in rotation (after preliminary enquiry with no penalty if declining invitation) (c4.1); in the tender documents full information should be given on fluctuations provisions, basic dates stated and the basic materials schedule prepared (c4.2); tender prices must not be disclosed to any other person (c5); holiday periods and design considerations must be taken into account in fixing times for tendering (c4.3).

Standing orders should regulate opening of tenders and notifying results. Where a subsidy is involved the consent of the Secretary of State is necessary before accepting other than the lowest tender. Passing over the lowest tenderer should not be necessary in selective tendering unless time and pricing considerations are involved. If eight weeks have elapsed since submission, confirmation of the tender before acceptance is recommended (c.5).

On the correction of errors in tendering, alternative one in paragraph 6.3 of the NJCC code is recommended (ie, confirm offer or withdraw tender). Alternative two is not suitable for local authorities (ie, amendment of tenders to correct errors).

Tenders should be invited from a limited number of contractors from a list of those expressing an interest in the project (the ad hoc list) or for any project for which they are suitable (the approved list). Detailed guidance is given on the preparation of lists. Approved lists should be revised at not less than 12 monthly intervals. (*Note*: A standard form of questionnaire for tenderers has been agreed between the NFBTE and local authority associations).

Negotiation
This can be advantageous and is acceptable if based on some preliminary price competition (eg, two stage tendering or continuation of original contract won in competition not more than two years before). Negotiation without competition should be the exception.

2.4.5 FIRM PRICE AND FLUCTUATION CONTRACTS *(CHAPTER 9)*
The contract period should generally determine which method is chosen. Adequate and thorough pre-planning is essential. In firm price contracts commitments are known, work in calculating fluctuations is saved and there is a greater incentive to complete on time.

Firm price fully pre-planned contracts should be for periods of up to one year. If an authority judges it to be in its own interests the period may be over a year.

In price fluctuation contracts tenderers are not exposed to unreasonable risks and tenders should reflect this. The contract period should normally exceed one year but in exceptional cases may be less. The two acceptable methods are traditional fluctuations (clause 39) with a percentage addition of up to 15 per cent under clause 39.8, or with formula adjustment (clause 40) with a 10 per cent non-adjustable element. The latter should normally be used for contracts between one and two years and either method if over two years. Formula adjustment is not suitable for certain labour intensive schemes. Clause 25 (extension of time) should operate in full.

A nominated subcontract under a main (fluctuation) contract should be let on the same basis as the main contract if it cannot be adequately pre-planned, if it will take more than 12 months, or if the programme dates are not known when tendering. In a main contract of over two years' duration on a clause 39 basis, nominated subcontracts may be let on the formula basis if nominated subcontractors prefer it and the client so decides.

2.4.6 CONTRACT CONDITIONS *(CHAPTER 10)*
The Standard Form of Building Contract should if possible be used unaltered, thus avoiding misunderstandings, additional expenses and claims. A reasonable contract period should be stated in tendering documents with phased handover in larger contracts. The documents may provide for alternative offers from tenderers based on their own contract period. A bonus for early completion is undesirable. Realistic liquidated damages encourage timely completion. Penalty clauses (as distinct from damages) must not be used.

Bond requirements, which are at the discretion of individual authorities, may be reduced where on selective tendering the financial standing and technical ability of tenderers have been investigated. Bonds should be dispensed with where it seems reasonable. If no bond is needed tenderers should be clearly told.

2.4.7 EXECUTION OF CONTRACT *(CHAPTER 11)*
The emphasis is on the need to ensure that the contract is efficiently run, that prescribed standards of materials and workmanship are met and that periodic

payments are properly calculated and promptly paid.

Once a contractor is appointed, planning and programming should start at once with the contractor given adequate time for pre-planning. Subsequent variations can disrupt the programme, delay completion and incur additional costs.

Attention is drawn to the Standard Form provisions on architect's instructions to (and through) the main contractor and to the need for frequent interchange of information, site meetings, etc.

The contractor must be clearly and comprehensively informed (by drawings, specifications, etc) on the requirements on quality of materials and workmanship for which he is responsible. Unsatisfactory work should be promptly notified and rectified.

Cost control on expenditure, including variations, should be carefully exercised. Contractors and subcontractors must be assured of a systematic cash flow. Dates for issuing and honouring certificates must be adhered to, interim certificates to be as fully valued as possible (eg, including variations).

An advisory group of chief officers will advise on the evolution of the code in the light of developments. Some revision will be necessary on the publication of the 1980 Editions of the Standard Form, the relevant clause numbers in which have been used above.

2.5 EEC tendering code

In 1971 there was an EEC directive (71/304) designed to enable contractors from member states to participate in public works contracts on equal terms with national contractors. Contracts must be advertised in the EEC official journal if they have an estimated value of at least one million units of account (EEC directive 71/305). From 1978 the European unit of account (£0.659 to one EUA) has raised the advertising limit from £415,000 to £660,000. Guidance on this directive is contained in DOE Circulars 59/73 and 102/76.

2.6 Tender documents

What are the tender documents? The brief answer is whatever the building owner or his architect provides. These generally range through the following:

(a) *Drawings only*
These should preferably be 1/100 scale with annotations, but larger scale drawings would be necessary for joinery, etc.

(b) *Drawings and specification*
Although larger architectural practices and local authorities produce clear and understandable specifications, there is no universally agreed standard

model form of specification. Many specifications are inadequate and some, in fact, irrelevant. The National Building Specification prepared by the RIBA with its detailed and precise descriptions has not yet been widely adopted by the profession. Tendering on a vague specification can only lead to trouble.

(c) *Drawings and bills of quantities*
Nowadays descriptive bills often make a specification unnecessary. Provisional quantities may, on occasions, be used necessitating remeasurement after the work has been executed and a special edition of the Standard Form is now available for use with approximate quantities (see section 3.11). The NFBTE 'Memorandum on Firm Price Tendering' (11.7.57) gave as complete documentation at the tendering stage:

(i) bills of quantities prepared in accordance with the principles of the SMM from the proposed contract drawings
(ii) a site plan
(iii) 1 : 100 scale drawings
(iv) all such details and drawings as may be necessary to enable the contractor to commence the works
(v) an adequate specification

In the Standard Form of Building Contract (with quantities) under clause 5, the contract drawings and contract bills together with the articles of agreement, appendix and the contract conditions are the contract documents. Although there is a reference to descriptive schedules, these cannot impose obligations beyond those set out in the contract documents. This applies equally to a master programme which is now the subject of an optional provision in clause 5.3 (see section 3.5.1). Bills of quantities are essential in projects of any size since they provide a uniform basis for pricing in competition, for evaluating tenders, and for valuations for payment, including variations (see NJCC Procedure Note No 12). Up to 1963 the NFBTE had a quantities rule under which members were not to tender in competition for work exceeding £8,000 in value unless quantities were provided. This rule was withdrawn after a reference to the Restrictive Practices Court. Quantities should unless otherwise stated in respect of any item, be prepared in accordance with the Standard Method of Measurement (SMM.) under clause 2 of the Standard Form (with quantities) which also states that the quality and quantity of the work shall be as set out in the contract bills. The bills cannot override or modify the contract conditions and errors should be corrected and valued as a variation of the contract. Any unspecified departure from the SMM or any errors in quantities or description in, or omissions from, the bills requiring correction would be treated as a variation and valued under clause 13.

The contractor can apply for an extension of time under clause 25 and under clause 26 submit a claim for loss and/or expense where regular progress has been affected by any discrepancy or divergence between contract drawings and/or contract bills which under clause 2.3 must be immediately notified in writing to the architect for an instruction (see also section 3.5.1). If the bills do not form part of the contract this protection is not available to the contractor.

2.7 Standard Method of Measurement

The first edition of the Standard Method of Measurement of Building Works made its appearance in 1922 under a joint committee represented by the Surveyors' Institution, the Quantity Surveyors' Association, the NFBTE and the IOB. It provided a uniform system of measurement of building work, giving unity to professional practice and confidence to builders tendering for contracts.

A second edition appeared in 1927, and in 1931 it was recognised by incorporation in the Standard Form of Building Contract. Other editions followed and a metric 5th edition was published in July 1968. The 6th edition was published in August 1978 and came into operation on 1 March 1979. A major revision is contemplated for a 7th edition which may be published some time in the 80s (see Appendix I for details of publications).

The introduction to the 6th edition stated that the SMM 'provides a uniform basis for measuring building works and embodies the essentials of good practice but more detailed information than is demanded by this document should be given where necessary in order to define the precise nature and extent of the required work'.

The SMM is divided into General Rules, Preliminaries and 20 work sections (C to X). It is very detailed and sets out the information which must be given under extensive trade divisions, what work shall be described, and in what terms.

Bills of quantities are prepared by quantity surveyors on this basis, information being 'taken off' drawings, etc, and 'worked up' into the bills in the prescribed form. This forms the basis for accurate and confident pricing by the contractor.

The 'junior' publication for the superstructure of small dwelling houses, the Code for the Measurement of Building Work in Small Dwellings, was first published in 1945. It is in an abbreviated form using the SMM as a basis. A third edition based on the 6th edition of the SMM has been prepared. This will be restricted to buildings of traditional construction up to two storeys and will not be appropriate to large schemes. There will be a mandatory rule limiting the use of the code to dwellings with an area not exceeding 130 square metres.

The Standing Joint Committee which publishes both documents is composed of members of the RICS and NFBTE. Rulings on agreed joint submissions on major issues in dispute may be given by this body.

2.8 Performance bonds

2.8.1 GENERAL CONSIDERATIONS

An interesting and important element in building contracting, which is not referred to in the conditions of contract or in tendering procedures, is the requirement on the contractor to provide a bond for the due performance of the contract if it should be considered necessary. This requirement should be indicated at the tendering stage.

The terms of the contract for the completion of the work are binding enough but if a contractor is unable to complete, the loss to the employer may be considerable. The purpose of the bond is to provide an independent and suitable guarantor who would be called upon to bear the cost up to the bond limit which would otherwise fall on the employer.

The two main sources of guarantees are banks and insurance companies. A bank's charges would very much depend on the contractor's financial standing. The fee might be based on a scale taken in conjunction with the risk. Collateral might possibly be sought in the form of title deeds, life policies and securities. The guaranteed amount could be counted in whole or in part against the overdraft limit. With insurance companies the premium is a percentage of the contract sum based on the contractor's financial position as disclosed by balance sheets. Counter indemnity may be sought and if this extends to the personal finances of directors it should be very carefully considered.

Contractors are generally opposed to bonds. Particularly in selective tendering, the contractor chosen should be financially sound. As work proceeds and retention increases, the risk shortens, particularly when there is always a fair percentage of work completed and unpaid at any one time. However, advocates of bonds, and in particular local authorities, point to possible losses if a contractor goes into liquidation.

Bonds cost money and, in the case of bank sureties, can restrict credit facilities. If a bond is contemplated, this could be covered by a provisional sum to be expended (ie, the cost added to the contract sum) if the successful tenderer is required to provide one. The limit should be 10 per cent of the contract sum with prompt release on practical completion.

Bonds may also be required from subcontractors and recent experience adds weight to the view that a bond from certain employers would be justified.

2.8.2 PERFORMANCE BONDS – LOCAL AUTHORITIES

In July 1976 (prior to the publication of Chapter 10 of the DOE Code of Procedures for Local Authority Housebuilding, referred to above in section 2.4.6) the Associations of Local Authorities after an approach by the NJCC issued the following guidance with their concurrence.

Each authority must make its own decisions on bond requirements. Where a bond is required:

(a) this should not normally exceed 10 per cent of the contract sum, full details of requirements to be given in tender documents

(b) release should be at practical completion (or earlier partial or sectional completion) unless for reasons stated release is delayed or is deferred until the end of the defects liability period

(c) the tenderer should be able to submit for reasonable approval, alternative sureties where the employer nominates the guarantor

(d) on receipt of proof that the cost of the bond has been met (eg, premium paid) this should be included in the earliest interim certificate.

3

THE STANDARD FORM OF BUILDING CONTRACT

3.1 Preliminary survey

This book concentrates on the Standard Form of Building Contract and the relevant subcontracts as published in January 1980. The Standard Form is the foundation of building contract conditions for all but Government work. Subcontract forms are related to and used in conjunction with it while some, like fixed fee, design and build and minor works forms are based upon it.

In considering the conditions of contract it must be borne in mind that although it is open to the parties to a contract to agree its terms within the framework of the law, the terms must be interpreted where appropriate, against the background of common and statute law and any relevant case law. Perhaps the most recent significant legislation affecting contract conditions is the Unfair Contract Terms Act 1977 which is summarised in Appendix H. In several instances provisions in the contract are 'without prejudice to any other rights and remedies' which the parties may possess.

A sound knowledge of the Standard Form by the professions, contractors, subcontractors and specialists is obviously essential both for the efficient conduct of building operations and for a proper understanding of the other forms of contract and their inter-relationship with the Standard Form.

Six editions of the Standard Form are available: the principal ones are the Private Edition, with or without quantities, and the Local Authorities Edition, with or without quantities. As far back as 1903 there was provision for quantities to form part of the contract where agreed (see Appendix E). A Standard Form with Bills of Approximate Quantities first appeared in October 1975 in both the private and local authorities editions.

Although the Standard Form is very widely used for private and local authority work where an architect is employed, the Government still uses its own conditions (GC/Works/1) for major new building work (see Chapter 9).

The Standard Form has proved successful in practice if judged by the relatively few disputes arising from its use. There are criticisms, however, which should not be ignored and as the form grows in complexity, scope for dispute widens.

Despite periodic major revisions, regular and necessary amendments have tended to make the form more cumbersome and complicated year by year. This may have caused a loss of cohesion and unity in the provisions; it has certainly reduced the number who really know and understand its terms. Architects, busy in their practices, and builders, engrossed in construction problems, are seldom really knowledgeable on the finer points of the contract unless their business organisations are big enough to afford a specialist in this field. The form has been criticised more than once in the courts and attention has been drawn to a certain obscurity and conflict in its lengthy provisions. Shortly after one judicial criticism in 1970 the title by which it was generally known, the RIBA Standard Form, was changed to the JCT Standard Form of Building Contract – a neat sharing of responsibility. Throughout this book the term Standard Form is used. The 1980 Private Edition (with quantities) published January 1980 is the form referred to, but the main differences between this and the Local Authorities Edition (with quantities) are indicated where appropriate. A separate section at the end of this chapter deals with the Standard Form with Bills of Approximate Quantities first published in October 1975. In the edition without quantities there are no contract bills, and the specification and schedule of rates become part of the contract documents. Clauses 2, 5 and 13 in particular reflect this difference from the 'with quantities' form.

The building contract conditions referred to in this book do not apply, as they stand, to Scotland. Scots law and practice requires modification and adaptation and this has resulted in separate agreements being prepared by the Scottish Building Contract Committee (SBCC) which is a constituent member of the JCT. The SBCC in turn has as constituent members the Royal Incorporation of Architects in Scotland, the Scottish Building Employers' Federation, the Scottish Branches of the RICS, FASS, CASEC and ACE and the Convention of Scottish Local Authorities.

This separate building contract incorporates the Standard Form conditions amended and modified as set out in the contract's Appendix. The provisions for determination by the employer are different because of Scots law on liquidation or bankruptcy. There are special conditions for payment for off-site materials and for arbitration by an 'arbiter'. Although a separate appendix provides for payment of VAT, this is in very similar terms to those in the Standard Form VAT Agreement. The attestation requirements are different and, reflecting Scots practice, there is provision for contracts let on a separate trades basis. The expression 'real or personal' is in Scots law 'heritable or moveable'.

Similar provision is made for the use of the nominated and non-nominated subcontract forms in Scotland.

The Standard Form has its critics. It has been suggested that employers, who order and pay for the work, have extensive commitments but very few powers under the contract and are not represented on the JCT, although the Confederation of British Industry now has 'observer' status. The Local Authority Associations, however, have been members of the JCT since 1957, and have, with the architects, watched over the interests of the employers. The more recent admission of the specialist organisations in 1966/7 may result, through the new JCT Nominated Subcontract Form, in a fairer balance between main and subcontracting interests (see Chapters 5 and 6).

Another criticism is one of practice rather than principle – the penchant of some to 'improve' the form by making their own amendments. It cannot be emphasised too strongly that if any of the conditions are considered capable of improvement, this should be represented through the appropriate constituent body of the JCT. All too often the 'improvers' fail to appreciate the inter-relationship of clauses, and many amendments are obscure giving rise to disputes. They are often incomprehensible to builders pricing the work. The value of agreed and understood standard conditions cannot be over-emphasised.

The regular amendments already referred to have become more complex and extensive in the past five years, and the possible reasons for this situation have been considered in Chapter 1.

3.2 The 1980 edition

These considerations resulted in the decision to publish the new 1980 edition of the Standard Form which takes the place of the 1963 edition with amendments (referred to in this book as the pre-1980 edition). The position has been made more complicated by the simultaneous publication of the first edition of the JCT Nominated Subcontract Forms which has necessitated radical but relevant changes to the Standard Form of main contract.

In preparing the 1980 edition the JCT dealt with four major matters:

(a) incorporation of all periodic amendments up to and including Amendment 15/78 (Social Security Pensions Act 1975 – fluctuations clause)
(b) major revisions of certain clauses including in particular those dealing with damages, extensions of time, nominated suppliers, work carried out by the employer, valuation of variations, payments and fluctuations, and arbitration
(c) a completely new clause on nominated sub-contractors needed to tie in with the new JCT Nominated Subcontract conditions published at the same time as the 1980 edition of the Standard Form
(d) the introduction of new provisions on selected subcontractors.

At the same time, the format was entirely altered, the contract now being in three parts with 40 clauses:

Part I General – clauses 1 to 34
Part II Nominated subcontractors and nominated suppliers – clauses 35 and 36
Part III Fluctuations – clauses 37 to 40

Most clause numbers are altered and the period to mid 1980 will be required for users of the conditions to accustom themselves to this change in addition to the major task of understanding the new provisions. Clause 1 is now devoted to interpretations and definitions. It is made clear that the Articles, Conditions and Appendix are to be read as a whole (clause 1.2).

The introduction of a definitions and interpretation clause (1) removes the need to define precisely on every occasion references to matters such as contract sum, interim certificates etc. Certain conditions however are referenced to the relevant clause eg, 'variations' in clause 13.1.

Repetition is also avoided by introducing a new clause 3 which states that any addition to, deduction from, or adjustment to the contract sum must immediately be taken into account in computing the next interim certificate.

In addition to this, the provisions on nomination (subcontractors and suppliers) are taken to new clauses 35 and 36. Clause 35 in the 1963 edition was the final one, dealing with arbitration. This provision now appears in Article 5 of the Articles of Agreement.

The fluctuations clause (formerly 31) is now at the end (clauses 37-40) and clause 37 briefly states that fluctuations are to be dealt with by one of three methods:

> 38 – contribution levy and tax fluctuations; *or*
> 39 – labour and materials cost and tax fluctuations; *or*
> 40 – use of price adjustment formula.

The method to be used is identified at tender stage in the Appendix with a stipulation that, if the alternative selected is not shown, clause 38 will apply. The alternatives are printed separately, the relevant provisions only to be incorporated in the contract when executed.

A new development at detail level is decimal numbering. Letters are now the exception but these are retained for alternative provisions (eg, fire etc insurance 22A, 22B, 22C). Letters are also used to avoid gaps where clauses apply to only one edition eg, fair wages 19A (Local Authorities edition). The clauses and subclauses are set out fully in a contents section.

To assist in understanding the new format and the relationship of clauses in the 1963 edition to those in the 1980 edition, Appendix D has been prepared as a broad general guide. It must be borne in mind that the individual clauses may be substantially changed in the new edition requiring detailed individual study. JCT Practice Notes on the pre-1980 edition still merit careful consideration. These are summarised in Appendix C. Special attention is drawn to the lengthy and valuable JCT Explanatory Memorandum on the 1980 edition of the Standard Form and the JCT Nominated Subcontract Forms (see Appendix I).

In preparing the second edition of this book it has been decided to include in this present chapter all the new provisions found in the 1980 edition of the Standard Form with the exception of fluctuations and those amendments and provisions directly related to the nomination of subcontractors. Chapter 6 brings together main contract provisions in clause 35 and the new JCT Nominated Subcontract Form's terms and procedures which are all inter-related. The earlier nominated subcontract provisions reviewed in Chapter 5 will continue to operate in existing main contracts for several years yet. When all contracts current at the time the new forms come into use run out, we may, in this respect at least, have a less complicated situation.

Because of their importance the fluctuations provisions are specially considered in Chapter 4.

3.3 The Articles of Agreement and the Appendix

The Articles of Agreement, to which the employer and contractor are the parties, set out the location and nature of the intended works, the contract sum, a reference to the contract drawings (identified by numbers), bills of quantities (signed by the parties for identification) and the conditions of contract. The architect and surveyor are named in the articles of agreement, but they are not parties to the contract, although the architect derives his extensive powers from its provisions. The term of the architect's appointment are a matter between him and the employer. He designs and supervises; the contractor constructs. Architects (and quantity surveyors) cannot plead that they were acting as quasi-arbitrators as a defence against claims for negligence in carrying out their functions.

The Articles of Agreement also make provision for the appointment of another architect or quantity surveyor by the employer, in the event of death or ceasing to act. In the Private Edition only, the contractor has the right of objection to the further appointment provided his reasons are considered sufficient by an arbitrator. The contractor is further protected under the Articles of Agreement by the provision that a subsequent architect cannot overrule the decisions, instructions, certificates, etc of his predecessor. If the employer refuses to appoint a successor the contractor can apply to the court.

A new recital (4) states that at the date of tender the status of the employer for the purposes of the statutory tax deduction scheme under the Finance (No 2) Act 1975 is as set out in the Appendix. (See later in this chapter (3.8.5) for details of the scheme and the provisions of clause 31.) This is essential information at the tendering stage which would thus be available through the Appendix details set out in the preliminaries in the bills.

As already indicated the arbitration provisions are now incorporated in the Articles of Agreement. These provisions and a consideration of the arbitration procedures are set out at the end of this chapter (3.10).

The Appendix which appears at the end of the conditions also sets out information related to, and necessary to complete, the conditions of each particular contract. It would be appropriate to consider this here, the items below which appear in the Appendix being reviewed in detail in this chapter. The references are to clause numbers in the contract.

- statutory tax deduction scheme – status of employer (recital 4 : clause 31)
- whether the joinder provisions in arbitration apply (article 5)
- defects liability period (6 months from practical completion certificate unless otherwise stated) (17.2)
- insurance cover (third party) (21.1.1)
- percentage to cover professional fees (22.A)

- date of possession (23.1)
- date for completion (25)
- liquidated and ascertained damages (24.2)
- period of delay for fire and other reasons (recommended three months and one month respectively) (determination by contractor) (28.1.3)
- period of interim certificates (one month unless otherwise stated) (30.1.3)
- retention percentage (if less than five) (30.4.1.1)
- period of final measurement (six months from practical completion certificate unless otherwise stated) (30.6.3.1)
- period for issue of final certificate (three months from end of defects liability period etc unless otherwise stated – six months maximum) (30.8)
- nominated subcontract work for which contractor desires to tender (35.2)
- fluctuations provisions (38, 39 or 40; alternatives not selected to be deleted; 38 to apply in absence of clear indication)
- percentage addition (38.7 or 39.8)
- formula rules data (40); base month, non-adjustable element percentage, work categories or work groups.

The page following the Articles of Agreement should be completed with the appropriate attestation clause, either under hand or under seal. Main contract and subcontracts should be executed similarly and special attention is drawn to the note later in this chapter on the Limitation Acts. (section 3.8.3)

3.4 Review of the contract's provisions

Basically this chapter endeavours to explain the essential points of a complicated contract. Authoritative books are available on special points of legal complexity. These works generally expound clause by clause. Here the subject is treated under the various stages of the execution of a contract.

As already indicated three major aspects of the contract are separately considered elsewhere, namely:

(1) firm price and variable price contracts *(Chapter 4)*
(2) nominated ('green' form) and non-nominated ('blue' form) subcontract forms *(Chapter 5)*
(3) nominated subcontract work under the provisions of JCT Nominated Subcontract Form NSC/4 etc and the related main contract clause 35 *(Chapter 6)*

There is a separate section at the end of this chapter on the Standard Form with Bills of Approximate Quantities first published in October 1975.

In addition to reviewing the three alternative provisions for fluctuations, Chapter

4 also contains information on incomes policy as it affected building contracts.

In a period of transition the subcontracting position needs to be spelled out carefully. Chapter 5 concentrates on the pre-1980 contract considerations affecting nominated subcontracts let under the terms of clause 27 (pre-1980 edition), ie, before the advent of the JCT Nominated Subcontract Form, but still current. The salient features of subcontracting and the distinction between nominated and non-nominated (domestic) subcontractors are also fully explained. Non-nominated subcontractors under clause 17 of the pre-1980 Standard Form are now covered by clause 19 of the new form which describes them as domestic subcontractors. The non-nominated subcontract provisions in the 'blue' form will therefore continue to apply with some necessary amendments. These are also reviewed in Chapter 5. The labour-only subcontract agreement is dealt with in Chapter 8.

The new Standard Form of Subcontract issued for the first time under the authority of the JCT which is reviewed in Chapter 6, will be used for nominated subcontracts under main contracts being carried out under the 1980 edition of the Standard Form. As already indicated, Chapter 6 brings together these provisions and the related terms of Clause 35 of the main contract.

3.5 Contract conditions and procedures

3.5.1 CONTRACT SCOPE AND DOCUMENTS

The conditions of contract (with the Articles and Appendix) the contract drawings and the contract bills are the contract documents and govern the work to be carried out. Materials and workmanship must be of the quality and standard specified and to the reasonable satisfaction of the architect where his approval is required (clause 2.1). Under clause 2.3 the contractor should at once notify the architect in writing and require a written instruction where any discrepancy or divergence between any of the following comes to his notice (see also clause 25.4.5 (extension of time) and clause 26.2.3 (loss and/or expense) – sections 3.6.9 and 13):

- (i) the contract drawings
- (ii) the contract bills
- (iii) an architect's instruction (other than a variation under clause 13)
- (iv) any contract documents issued under clause 5
- (v) details on setting out, etc under clause 7

The architect (the employer in the Local Authorities edition) is the custodian of the contract drawings and bills to be available at all reasonable times for inspection by the contractor (clause 5.1). Under clause 5.2 the contractor is entitled, free of charge on the signing of the contract, to a certified copy of the contract documents and two further copies of the drawings and two copies of the blank bills. Clause 5.5 states that a copy of the contract drawings, and further details, the blank bills, descriptive schedules, etc must be kept on the site. Under clause 5.3 two copies of descriptive schedules should be provided where necessary but these are not contract documents and cannot impose any obligations not contained in the contract documents themselves. Although the architect decides if descriptive schedules, etc are necessary, the contractor is justified in asking for all the necessary information for the satisfactory execution of the work. There is now an optional provision in clause 5.3 requiring the contractor to supply two copies of his master programme for the works and to up-date within 14 days of any decision under clause 25 (extensions of time). This does not impose additional contractual obligations, but it should be available on site.

Under clause 5.4 further drawings or details, in duplicate, to explain or amplify the contract drawings or to enable the contractor to carry out the work must be issued 'when necessary' – presumably when the contractor reasonably requires them (see clauses 25.4.6 and 26.2.1). The architect may, under clause 5.6, require the return of drawings, descriptive schedules, etc once final payment is made.

Clause 5.7 points out the restriction on the use of documents and that priced bills are confidential and must only be used for the purpose of the contract. This is a matter to which contractors rightly attach great importance.

The contract bills set out the quality and quantity of the work included in the contract sum (clause 14.1). Unless otherwise stated in respect of any item the bills are to have been prepared in accordance with the Standard Method of Measurement (SMM) 6th edition (clause 2.2.2). No provision in the bills can override or alter the contract conditions including the Articles and the Appendix (clause 2.2.1). If a contractor finds a conflict between the bills and the contract conditions, he should refer the point at once to the architect.

Any departure from the SMM, as provided for in clause 2.2.2, or any errors in the bills in description, quantity etc, should be promptly notified to the architect so that he may correct them, the correction to be treated as a variation requiring an instruction under clause 13.2 and valued accordingly (clause 2.2.2). After the contract is made, errors in pricing, etc which come to light are deemed to have been accepted by the parties (clause 14.2).

3.5.2 ARCHITECT'S INSTRUCTIONS

Clause 2.1 states that the contractor undertakes to carry out and complete the work in accordance with the contract documents, the materials and workmanship to be of the quality and standard specified, or to the reasonable satisfaction of the architect where his approval is required (prompt written confirmation of this should be sought by the contractor). During the currency of the contract the architect may issue instructions, but his power to do so is derived solely from the contract. If the contractor thinks the architect is going beyond his powers in issuing a particular instruction, he should ask him, in writing, for a note of the clause under which it is issued (clause 4.2). Once this is received, the contractor, if satisfied, must at once comply with the instruction, any variations to be valued under clause 13. If on the other hand the contractor challenges the reply, the next step is to give written notice at once of immediate arbitration. If the architect 'forthwith' (say within seven days) fails to state the clause under which the instruction is issued, it may, under clause 4.2, be ignored by the contractor.

An important new point on complying with instructions relates to any increase in the restrictions an employer has set out in the bills on access, limitation of space or hours, or the order of work, requiring a variation under clause 13.1.2. A contractor, if he has notified to the architect in writing a reasonable objection to a significant increase in restrictions, may refuse to comply. Any dispute on whether such a refusal is reasonable is immediately referable to arbitration (see 'variations' – section 3.6.4).

This apart, immediate compliance with architect's instructions is vitally important. Failure by the contractor to comply within seven days after a written notice requiring this, may result in someone else being employed to do the work. Under clause 4.1 the cost is recoverable from the contractor and may be deducted from monies due to him under the contract.

Oral instructions (ie not in writing) should not be acted on.

Written instructions are essential and a period of seven days is given for the architect to confirm oral instructions in writing. Should they be given and not confirmed, the contractor should confirm in writing to the architect within seven days. The architect then has a further seven days in which to dissent. These cumbersome but essential procedures can be reduced if the architect signs a memo confirming instructions before leaving the site. It can avoid later argument if some indication of any additional cost is promptly given.

Under clause 4.3 there is provision for a contractor acting on an instruction other than in writing. It may be made effective from that date by confirmation at any time prior to the issue of the final certificate, but only in the most exceptional circumstances should the contractor proceed without adequate evidence in writing. Instructions may include postponement of work (clause 23.2). The architect under clause 11 has a right of access at all reasonable times to the work, workshops, etc of the main contractor who must ensure a similar right in respect of all his subcontractors.

3.5.3 STATUTORY OBLIGATIONS

Clause 6 was extended in July 1975 and again in July 1976. The situation when the contract instructions are at variance with statutory obligations is covered as follows:

(a) The contractor shall comply with, and give notices required by, legislation, regulations, or byelaws in regard to any local authority or statutory undertaker concerned with the works (6.1.1).

(b) If the contractor finds a divergence between contract documents or instructions, and statutory requirements, the contractor must give the architect immediate written notice of the divergence. On receiving the notice or otherwise discovering the divergence, the architect must issue his instructions within seven days (6.1.3), any variations to be valued under clause 13. If this procedure is followed the contractor will not be liable to the employer in complying with the contract documents or a variation, if the works fail to meet statutory requirements (6.1.5).

(c) If emergency action is required under statutory obligations because of a divergence before the contractor receives an instruction, then the architect must be informed at once of the necessary steps taken, the work to be regarded as carried out under an instruction and dealt with as a variation (6.1.4).

There is the long standing provision that fees and charges (including rates or taxes) legally demandable under any act or any byelaws, etc of a local authority or statutory undertaker (eg, rates on site huts, etc) are to be met by the contractor and added to the contract sum unless covered by a bill item, a provisional sum, or the costs result from a nomination for work or materials (6.2). It is made clear in clause 6.3 that none of the provisions of clause 35 (nominated subcontractors) or clause 19 (sublet work) apply where PC items in the bills, or arising on the expenditure of provisional sums in respect of fees and charges under this clause relate to any authority or statutory undertaker in respect of statutory obligations ie, they are not subcontractors. Amounts paid are to be added to the contract sum and included in the next interim certificate (not subject to retention – see section 3.8.2).

3.5.4 CLERK OF WORKS

The clerk of works, appointed by the employer, is only an inspector under the architect. If he gives directions they must be within the scope of the architect's powers and these are only effective if confirmed by the architect in writing within two working days. Except in emergency work (eg, under clause 6) the contractor should await the necessary architect's instruction before complying with a clerk of works' direction, particularly if he considers it onerous or unreasonable (clause 12). The contract provisions apart, a good relationship between a reasonable clerk of works and a responsive and responsible contractor (or his agent) is one of the elements of a successful contract. The contractor must in any event 'afford every reasonable facility' for the performance of the clerk of works' duties.

3.6 Carrying out the contract

3.6.1 POSSESSION, COMPLETION AND DAMAGES

Under clause 23.1 provision is made in the Appendix for the insertion of the date of possesssion. If possession is not given when specified, this would be a breach of the contract conditions. In addition failure to provide agreed access, etc could give grounds for an extension of time (clause 25.4.12) and a loss and/or expense claim (clause 26.2.6). If the architect, in writing, postpones the work under clause 23.2 the contractor may claim under clause 26.2.5 if direct loss and/or expense occurs, in addition to an extension of time under clause 25.4.5 and grounds for determination under clause 28.1.3.4.

The contractor must proceed regularly and diligently with the works (see clause 27.1.2 – grounds for determination). Failure to complete on time, as extended under clause 25, may result in the enforcement of damages for non-completion under clause 24.

When in the opinion of the architect practical completion is achieved he must forthwith issue his certificate of practical completion identifying the date (clause 17.1).

There are three essential requirements in clause 24 in applying damages for failure to complete by the completion date.

(a) The architect must under clause 24.1 issue a certificate that the contractor has failed to complete the works by the completion date defined ie, the date originally stated in the Appendix as extended under clauses 25 and 33 (war damage) (ie, it is a condition for the recovery of damages). This would be a confirmation of the completion date under the clause 25 procedures.

(b) The employer, before the issue of the final certificate, must notify the contractor in writing of his intention to invoke the damages clause. The employer has the discretion as to whether or not to enforce the provisions on liquidated damages.

(c) The contractor must then pay or allow damages in whole or in part as the employer may decide, at the rate set out in the Appendix for the period between the completion date (see (a) above) and practical completion (certified under clause 17). Normally damages would be deducted from certificates (both interim and final). Otherwise it would be recoverable as a debt.

If the architect, not later than 12 weeks after the stated practical completion date, certifies under clause 25.3.3 a later completion date, any excess damages deducted must be repaid by the employer (clause 24.2).

There is no implied right to a bonus for early completion, and the employer is not bound to take over the works until the contract completion date.

When damages are specified, these are enforceable provided they do not amount to a penalty, so it is important that the contractor should satisfy himself at tendering stage that they are a reasonable pre-estimate. The employer should fix damages at a level which would compensate him for loss of rents or production, storage, alternative accommodation, interest charges and any additional costs where the contract overruns. He is not required to prove before deduction that the pre-estimate was in fact correct nor can the contractor claim that the actual loss was less than the agreed damages. When the employer at the outset does not propose to avail himself of the contract's provision for damages, this could be clearly indicated in the Appendix but as outlined above the employer can waive prescribed damages in whole or part before the final certificate is issued.

Clause 24 is closely related to the provisions on extension of time under clause 25 (see section 3.6.9 which should be read in conjunction with this section).

3.6.2 SETTING OUT

To enable the contractor to set out the work, accurately dimensioned drawings should be supplied by the architect, who should also determine the levels required. Any inaccuracy in setting out is the responsibility of the contractor (including liability for trespass) unless the architect otherwise instructs (clause 7). Delay in providing this information could give rise to a claim for an extension of time under clause 25.4.6 or for loss and/or expense under clause 26.2.1.

3.6.3 MATERIALS AND WORKMANSHIP

Tests

Opening up covered work for inspection or tests of materials or goods, may be ordered by the architect. Under clause 8.3 the cost is to be added to the contract sum (not subject to retention) unless provided for in the bills, or unless the workmanship, materials, etc are not in accordance with the contract. (See also extensions of time clause 25.4.5.2 and loss and/or expense clause 26.2.2 – sections 3.6.9 and 3.6.13.)

Materials, etc

Under clause 8.1 materials, etc, as far as procurable, must be as described in the bills. If materials are no longer procurable, an architect's instruction should be sought at once. Clause 8.2 states that substantiating vouchers should be provided if required.

Under clause 8.5 the architect may, where reasonable, order the exclusion of any person from the works and, under clause 8.4, he may require the removal of unacceptable work, materials, etc. Access to the works and workshops of main contractor, and subcontractors must be available to the architect at all reasonable times under clause 11.

Materials – unfixed

Materials on the site not yet fixed cannot be removed without the written consent of the architect. Under clause 16.1, materials when paid for, become the property of the employer, but the contractor remains responsible for loss or damage (but see 'fire' clauses 22B and 22C).

Under clause 30.3 off-site materials ready for incorporation may, at the discretion of the architect and subject to clearly defined conditions (including insurance for full value under clause 22 until delivered to the site), be included in interim certificates. (See section 3.8.2) When paid for, they become the property of the employer. Except for use on the works they are not to be removed and, under clause 16.2, the contractor is responsible for loss or damage and for the cost of storage, handling and insurance prior to delivery to site.

Following the 'Romalpa' case, increasing use was made of retention of title provisions in contracts for the sale of goods – eg, the property in the goods does not pass until paid for. Since clause 16 provides that the property in on or off-site materials passes to the employer when amounts due in certificates in which those materials are included have been paid, the contractor may well be required before certification to show that he has a good title to the materials in question. (*Note*: proof of title is not required in clause 30.2 in respect of on-site materials, although it is in clause 30.3 for off-site materials.) However, once materials are incorporated into the works they become the property of the employer. In August 1978 the JCT considered that the risk to the employer was so slight in respect of on-site materials as not to require any adjustment to the contract's terms (eg, a risk could arise on a contractor's insolvency between the date of payment by the employer for the goods and their incorporation in the works).

It should be specially noted that a condition of nomination of a supplier under clause 36 is that there should be no retention of title condition (see section 3.6.6).

3.6.4 VARIATIONS

The architect has power to order variations within the terms of the contract, but these variations do not vitiate it (clause 13.2). Variations by the contractor without an instruction may be subsequently sanctioned in writing under this clause, but a prior written instruction is always desirable. Although under the contract the variations which can be ordered are very widely defined, the contractor might challenge those which completely altered the nature or conduct of the contract. (see in particular clause 13.1.2 referred to below) Any nominations of subcontractors or suppliers on instructions on the expenditure of provisional sums are dealt with under clauses 35 and 36. There is no longer a reference in this clause to instructions on PC sums (see section 3.6.5). The ways in which a subcontractor may now be nominated are set out in clause 35 (section 6.3.1).

Scope

Variations under clause 13.1.1 are as follows:

(a) the alteration or modification of design, quality or quantity of the contract works
(b) the addition, omission or substitution of any work
(c) the alteration of the kind or standard of materials or goods to be used
(d) the removal of work materials or goods, other than that not in accordance with the contract – see clause 8.4.

In addition there is a new provision in clause 13.1.2 which brings within the definition of a variation, and thus requiring valuation, subsequent changes in those employer's requirements which the Standard Method of Measurement B.8 requires to be set out in the bills in respect of the following: addition to, alteration or omission of any obligations and restrictions on site access, limitation of working hours or space, or the execution or completion of the work in any specific order.

If a contractor has reasonable objection to any changes in requirements and so notifies the architect in writing then to that extent he need not comply with the instruction and clause 4.1 provides accordingly. Any dispute on such an objection could be referred to immediate arbitration.

Under clause 13.1.3 a variation cannot be used to nominate a subcontractor for work measured and priced in the contract bills by the contractor (but see clause 35.1.4).

Under clause 13.3 an architect may issue his instructions on the expenditure of provisional sums in the contract bills, and also in a subcontract.

There are also certain other clauses of the contract under which instructions are to be regarded as a variation under clause 13 and these can usefully be listed here:

(i) where work is varied by an instruction because statutory obligations diverge from contract documents or instructions or emergency action is required (clause 6.1)
(ii) any correction of an error or omission in the bills (clause 2.2)
(iii) making good fire etc. loss (clauses 22B or 22C)
(iv) work partially completed when the contractor determines (clause 28.2.2)
(v) work under war damage clause 33.1.4.

Valuation of variations

The provisions of clause 11 of the pre-1980 Standard Form on the valuation of variations on occasion gave rise to argument. In brief the architect has very extensive powers to order variations but the contractor is entitled to be fully re-imbursed for the costs involved. There was a tendency to apply bill rates where appropriate to the exclusion of other costs incurred and although clause 11.6 gave the opportunity of some relief contractors sometimes felt they were being under-paid for variations. Adequate recompense is essential, no more and no less. The provisions should not afford relief for low initial pricing. The Joint Contracts Tribunal have revised and clarified the provisions which appear in clause 13 in the 1980 edition. In valuing variations it is clearly intended that significant changes in quantity must now be taken into account along with appropriate adjustments for preliminary items and lump sum elements in pricing and the effect of a variation on other work. The loss and/or expense element under former clause 11.6 is now dealt with as an additional item in clause 26 (formerly clause 24) (see section 3.6.13).

Having outlined changes in general terms we must now look at the rules for the valuation of variations as set out in clause 13.5.

Rules for valuation

Variations required or subsequently sanctioned in writing by the architect and all work executed by the contractor under a provisional sum must be valued by the quantity surveyor. The contractor has the right to be present, where it is necessary to measure the work, to take notes and measurements (clause 13.6). Where a contractor tenders for and carries out work under a PC sum on the expenditure of a provisional sum under this clause, which would otherwise be work reserved for a nominated subcontractor, (clause 35.2), valuation is to be on the basis of the contractor's accepted tender. Under clause 13.4 the valuation of variations and of provisional sum work carried out by a nominated subcontractor under clause 13.3.2 must be in accordance with his subcontract (NSC/4 clauses 16 and 17), unless otherwise agreed.

It is much more satisfactory to agree a price or a basis for pricing a variation with the employer before it is carried out but if this is not possible valuations are to be made in accordance with subclause 5 which provides as follows:

> To the extent that 'The Valuation' is for additional or substituted work which can be properly valued by measurement, valuation shall be in accordance with the following rules under clause 13.5.1.
>
> (i) Work which does not significantly change the quantity and is of a similar character, executed under similar conditions – the prices in the bills to be used (also applies generally to omissions – 13.5.2).
> (ii) Work of a similar character but not executed under similar conditions and/or with significant changes in quantity – the prices in the bills to be a basis for valuation with a fair allowance for differences in conditions and/or quantity.
> (iii) Work not of a similar character – valued at fair rates and prices.

It is stipulated in clause 13.5.3. that in these valuations the principles of measurement governing the contract bills should be used and that allowances be made for any percentage or lump sum adjustments in the bills and where appropriate for adjustment of preliminaries as defined in Section B of the Standard Method of Measurement (6th edition). (*Note*: This will now reflect more fully the true cost of variations to the contractor).

Where additional or substituted work cannot properly be valued by measurement, the valuation shall be as follows under clause 13.5.4:

> (i) prime cost as defined in the edition of the RICS/NFBTE Definition of Prime Cost of Daywork under a Building Contract which is current at the date of tender plus percentage additions set out by the contractor in the contract bills (see section 8.7 for information on the RICS/NFBTE Definition)
> (ii) in subcontracting trades – any specially agreed prime cost definitions current at the date of tender plus percentage additions set out by the contractor in the contract bills (see also NSC/4 clause 16 – section 6.4.6).

Since the value of variations is to be adjusted in the contract sum (clause 13.7) and under clause 3 included once computed in the next interim certificate it is essential that the contractor's labour, plant and materials vouchers establishing prime cost should be sent to the architect by the end of the following week for prompt payment.

There remain two points – the effect of a variation on other work in the contract and its effect on progress etc.

Clause 13.5.5 states that if a variation or the expenditure of a provisional sum valued under clause 13 substantially changes the conditions under which other work in the contract is executed, then that other work shall 'be treated as if it had been the subject of an instruction requiring a variation', to be valued accordingly.

If the application of the provisions outlined above does not result in a reasonable valuation, or if a variation or the expenditure of a provisional sum does not entail additions or omissions to the work, then a fair valuation has to be made. This wide ranging provision should ensure adequate reimbursement. However it is clearly stated that no allowance shall be made under this clause (13.5.6) for any effect upon the regular progress of the works, or for any other direct loss and/or expense for which the contractor could claim under any other clause. As has already been indicated clause 26.2.7 now includes grounds for a claim for direct loss and/or expense where regular progress has been, or is likely to be affected by architect's instructions under clause 13 requiring a variation or in regard to the expenditure of a provisional sum.

3.6.5 PRIME COST AND PROVISIONAL SUMS

It is important to note the distinction between prime cost (PC) and provisional sums. These are not defined in the conditions but the following definitions appear in the Standard Method of Measurement (6th edition – A.8):

- *Provisional sum*: a sum provided for work or for costs which cannot be entirely foreseen, defined or detailed at the time the tendering documents are issued.

- *Prime cost sum*: a sum provided for work or services to be executed by a nominated subcontractor, a statutory authority or a public undertaking or for materials or goods to be obtained from a nominated supplier. Such sum shall be deemed to be exclusive of any profit required by the general contractor and provision shall be made for the addition thereof.

Provisional sums being amounts included in the bills to cover work, etc which cannot be entirely foreseen, defined or detailed in the tendering documents, may, under clause 13.3, be expended on an architect's instruction both in regard to main and subcontracts. If the contractor is required to carry out the work it must be valued as a variation under clause 13.4. (See also NSC/4, clauses 16 and 17.)

A prime cost sum in the bills is one means whereby an architect, by an instruction, may nominate a subcontractor under clause 35.1 or a supplier under clause 36.1. A provisional sum under clause 13.3 may also be expended by the architect, by nominating a subcontractor or supplier to be covered by the provisions of either clause 35 or 36. A subcontractor cannot be nominated by means of a variation to carry out work already set out and priced in the bills for execution by the contractor nor, subject to clause 35.2, can a contractor be required to execute nominated subcontractor's work (for full information on nominated suppliers see section 3.6.6 and for nominated subcontractors section 6.3.1).

Since PC sums are exclusive of the main contractor's profit, this should be priced as a percentage of the sums shown in the bills, for adjustment pro rata in the final computation of the contract sum under clause 30.6.2.10.

It should be noted that although a PC or provisional sum may relate to statutory undertakers, etc under clause 6.3 they do not become nominated subcontractors or suppliers in respect of statutory duties under clauses 35 or 36 but are dealt with under clause 6.

Since under clause 35 a subcontractor may be nominated other than by means of PC sum, it is stipulated that SMM B9.1 is to that extent departed from.

3.6.6 NOMINATED SUPPLIERS

Clause 36 (nominated suppliers) (formerly 28) has been substantially revised and extended in the 1980 edition. It now also covers the specifying of a sole supplier from whom alone the contractor must obtain certain materials or goods, or where the materials or goods specified can only be obtained from one supplier.

To come within the contract definition of a nominated supplier who is to supply materials and goods to be fixed by the contractor nomination is essential under a prime cost sum on which the architect issues instructions (clause 36.2).

Under clause 36.1 prime cost sums can be created as follows:

(i) *by inclusion in the contract bills*
the supplier will either be named in the bills or subsequently named in an architect's instruction under clause 36.2

(ii) *where there is a sole supplier*
  (a) on the expenditure of a provisional sum – this should be made the subject of a PC sum the sole supplier being nominated.
  (b) in a variation – if as a result of a variation under clause 13.2 certain materials and goods are specified for which there is a sole supplier, this should be made the subject of a PC sum in the instruction, the sole supplier being nominated under clause 36.

Only sole suppliers who are dealt with in one of these ways will be deemed to be nominated by the architect under this clause. Materials and goods specified by the architect in the bills must be the subject of a PC sum if the supplier named, even if a sole supplier, is to be a nominated supplier. This point must be specially watched by contractors.

In ascertaining the costs to be set against a PC sum in respect of payments to nominated suppliers, all discounts (other than the cash discount) are to be deducted. Costs include statutory taxes and duties not otherwise recoverable (other than VAT which the contractor can treat as input tax), the net cost of packing, carriage and delivery and any contractual price adjustment (eg, price fluctuation) less trade discounts. It could happen that in obtaining these materials the contractor, in the opinion of the architect, properly incurred additional expense which would not be otherwise reimbursed (eg, necessary delivery costs not provided for in the quotation). This, if approved, could be added to the contract sum and under clause 3 included in the next interim certificate since it is outside the terms of the supply contract (clause 36.3.2).

The conditions of sale are most important and unless the architect and contractor otherwise agree, a supplier shall not be nominated under clause 36.4 who will not enter into a contract of sale providing, *inter alia*, for the following:

(a) Materials or goods to be of the quality and standard specified and to the reasonable satisfaction of the architect where required (see clause 2.1 for contractor's obligations).

(b) Replacement of goods where defects appear, up to the last day of the Defects Liability Period under the contract and the expenses reasonably incurred by the contractor as a direct consequence of this to be borne by the supplier (eg, labour costs in removing and refixing). This liability does not extend to defective materials which have been used or fixed and the defects were such as a reasonable examination would have revealed before fixing. Furthermore defects must be those arising from defective workmanship or materials and not from improper storage by the contractor or misuse or neglect by the contractor, architect, employer or any other person for whom supplier is not responsible.

(c) Delivery must be commenced and carried out and completed in accordance with the agreed programme or, in the absence of this, the reasonable directions of the contractor. There will be no obligation to deliver after the contractor's employment has been determined, unless the goods are fully paid for.

(d) Ownership of goods to pass on delivery to the contractor whether paid for in full or not (ie, this new stipulation means that there must be no retention of title provision in the conditions of sale. If there is, the nomination is invalid through failure to comply with the requirements of this clause.)

(e) Payment must be made in full within 30 days of the end of month of delivery, when the supplier will allow the contractor 5 per cent cash discount on all payments.

(f) If in the event of a dispute or difference between the contractor and the supplier being substantially the same as one already before an arbitrator under the main contract, the issue shall be referred to the arbitrator for a final and binding award, if the joinder provisions apply.

(g) No special condition of the supplier shall override or modify the conditions for nomination under this clause.

If a nominated supplier seeks to limit or exclude his liability to the contractor in any way outside the terms listed above, the architect must specifically approve this restriction in writing before the contractor is required to enter into a contract with the nominated supplier. When this happens, the liability of the contractor to the employer will be similarly restricted. This provision does not enable an architect to nominate a supplier who will not fully comply with the above conditions of sale (clause 36.5).

The contractor's right of objection to a nominated supplier is not so extensive as that to a nominated subcontractor, being virtually confined to points under the conditions of sale.

On the revision of this clause the Joint Contracts Tribunal prepared a new Standard Form of Tender for Nominated Suppliers. Its use is not mandatory but it could be used with considerable advantage when the architect nominates suppliers under clause 36 and the Tribunal recommends its use. This new form of tender is reviewed in section 7.4.

3.6.7 SUBLETTING AND ASSIGNMENT

Clause 19 of the 1980 edition adds to the provisions on assignment and subletting, a new class of 'selected' subcontractor.

Subclause 1 provides that the written consent of the other party is necessary for assignment of the contract.

Permission to sublet any portion of the works requires the written consent of the architect which shall not be unreasonably withheld (clause 19.2).

Failure to comply with either requirement gives grounds for determination under clause 27.1.4.

A new subclause 3 deals with certain work measured or otherwise described in the bills for pricing by the contractor, but which the architect requires to be carried out by persons listed in the bills from whom the contractor selects. The contractor's position is safeguarded by requiring at least three persons to be listed before the subcontract is let. If less than three are able and willing to carry out the work, the list may be brought up to this number at least, by the addition of names by the employer or the contractor, with the agreement of the other. Additional names may also be added to the list by mutual consent (all additions to be initialled by the parties). If less than three names are available for selection, the contractor may carry out the work himself or sublet under clause 19.2. The architect cannot then nominate (see clause 13.1.3).

It is important to stress that despite 'selection' this is not a nomination and the non-nominated subcontract conditions would apply. In work sublet under subclauses 2 and 3, there must be a stipulation that the subcontract will be terminated immediately on determination of the main contractor's employment under the contract (clause 19.4).

Apart from the opportunity to tender for work otherwise reserved for a nominated subcontractor under clause 35.2, a contractor cannot be required to supply and fix materials and goods, or execute work which is to be carried out by a nominated subcontractor (clause 19.5).

All subcontractors other than those nominated under clause 35 will now be known as 'domestic subcontractors' and described in all JCT documents as such.

3.6.8 FAIR WAGES

Clause 19A appears only in the Local Authorities edition. In brief, it requires the building contractor to pay wage rates and observe hours and conditions not less favourable than those laid down in the working rule agreement. Overtime permits should be obtained and the contractor should pay prescribed overtime rates where appropriate. Freedom to join trade unions, observance of these requirements by subcontractors, the display of the conditions of clause 19A on the site, etc, and the availability of wages records for inspection by the employer, are further duties laid on the contractor.

Issues arising under this clause, unless otherwise settled are to be referred to an independent tribunal through the Secretary of State for Employment and are specifically excluded from arbitration under the Articles of Agreement (5).

3.6.9 DELAY AND EXTENSIONS OF TIME

The provisions on extensions of time under clause 25 are closely related to the enforcement of liquidated and ascertained damages under clause 24. In brief the date for completion set out in the Appendix to the contract may be extended in prescribed circumstances and for defined reasons in accordance with procedures set out in clause 25. There are two important definitions in clause 1:

- 'completion date', the date stated in the Appendix as extended under clauses 25 or 33
- 'relevant event', one or more of the events (giving cause for an extension of time) referred to in clause 25.4

If a contract is not finished by the completion date the architect must issue a certificate that the contractor has failed to complete by the date specified and as defined under clause 25 (clause 24.1). As already explained, damages at the rate stated in the Appendix are then enforceable against the contractor for the period between the defined completion date and the actual date of completion. The employer must give notice before the issue of the final certificate of his intention to enforce damages. It is open to him to decide whether to enforce damages in whole or in part. Normally deductions would be made from monies due under the contract (clause 24) (see section 3.6.1). The arrangements for extension of time have three facets:

(i) The contractor's prompt notification of the circumstances and cause of delays or possible delays and their likely extent and effect enables the architect to review progress and decide what course, if any, he should take to minimise the effects of the delays (eg, changes in his design and specification or in the contractor's programming for which necessary recompense under an architect's instruction would be sought).

(ii) The contractor, by following the required procedures and obtaining an extension of time, has the date for the operation of clause 24 (damages) put forward.

(iii) The correct operation of all the requirements of the clause preserves the employer's right to damages for late completion.

It could be contended that as the result of an architect's failure to deal with all requests for an extension of time by the final dates prescribed, time was at large and the claim for damages lost. As long as architect's decisions are awaited on extensions of time under an unamended clause 25, provisions restricting the operation of clauses 38 and 39 (fluctuations) and 40 ('freezing of indices') do not come into effect (see sections 4.2.5 and 4.5). Where restriction or freezing is in operation the provision for immediate arbitration on issues under clause 25 is important.

Procedures

The 1980 revision of clause 25 (formerly clause 23) distinguishes between delays to the progress of the works for defined 'relevant events' (see below) which would give rise to a possible extension of time, and delays for other reasons which would not.

As soon as it becomes reasonably apparent that progress is being or it likely to be delayed for whatever reason the contractor must at once give written notice to the architect setting out the circumstances and cause of the delay. Where this delay is due to a 'relevant event' (giving a basis for an extension of time) this must be identified in the notice and the contractor must also in respect of each event give in the notice, or as soon as possible thereafter, written particulars of the expected effects and likely extent of the delay affecting the completion date. Where there is a reference to a nominated subcontractor in any notice, the contractor must send him a copy at once. The architect must be kept up-to-date by further written notices where necessary, or as required, including information on material changes in particulars or estimates (clause 25.2).

This clause previously required the architect to make an extension 'as soon as he was able to estimate the length of the delay'. If decisions were delayed (sometimes until after the completion of the works) this could be a major disadvantage to the contractor since he would be lacking essential information for any reprogramming and reorganisation of his contract which he might consider necessary, and be unable to estimate the financial repercussions.

Under the revised clause (25.3) there is a time limit of 12 weeks from the contractor's notice and particulars and estimates of relevant events, by which the architect must, 'if reasonably practicable' fix in writing to the contractor such later completion date as he considers fair and reasonable, if he is satisfied that completion of the works is likely to be delayed beyond the completion date ie, the date in the Appendix – as already extended. If a notice is given within the 12 weeks before the completion date, the decision must be made not later than that date. If there is a contractor's master programme under clause 5.3 (see section 3.5.1) this must be updated and notified to the architect whenever extensions of time are granted. Both parties must play their part in observing this timetable.

In granting extensions and fixing a later date for completion the architect will on each occasion consider all the 'relevant events' notified in writing by the contractor and take into account any omissions under clause 13.2 since fixing the previous completion date. In notifying his written decision to the contractor the architect must state the notified 'relevant event' taken into account and any variations requiring omissions since the previous decision but he is not required to apportion time to each event with the exception of those giving rise to monetary claims. Thus, in respect of certain relevant events under clause 25.4.5. a claim could arise under clause 34 (antiquities); and under clause 26.1 in respect of clauses 25.4.5, 25.4.6, 25.4.8 and 25.4.12. These clauses have been framed so that the architect must identify the actual time extension for each relevant event where necessary for claim purposes.

There is a provision requiring the architect to review finally the position, taking into account all relevant events, whether notified or not, as well as omissions since the last decision on extension, but this final decision on the completion date must be given by the architect within 12 weeks after the date of practical completion and conveyed in writing to the contractor (clause 25.3.3). This date may be the one previously fixed; may be a later one having regard to the circumstances and all the relevant events, whether notified or not; or may be an earlier one if omissions have been ordered since the last extension. However the architect cannot under clause 25.3.6 fix an earlier completion date than that stated in the Appendix. The Completion Date is therefore the date stated in the Appendix as varied by the architect under this clause. It is from this date that damages for non-completion will operate. (The remote event of war damage also warrants extensions under clause 33.1.3.) The decisions of the architect fixing a later completion date must be notified in writing to every nominated subcontractor as well as to the main contractor (clause 25.3.5).

There is the usual proviso that the contractor must use his best endeavours to prevent delay and to do all that may reasonably be required by the architect to proceed with the works. (The contractor is not thereby required to incur additional costs unless covered by an architect's instruction.)

Relevant events

There are 12 relevant events in clause 25.4 which could cause delay and give grounds for requests for extension of time. These can be divided for the sake of clarity into the following reasons (identified by sub-clause references).

(i) *External causes*

25.4.1 *force majeure* (eg, Act of God, war, floods, national emergency, etc)

25.4.2 exceptionally adverse weather conditions (too hot as well as too cold?)

25.4.3 fire, etc loss or damage under clauses 22A, B or C (ie "Clause 22 perils")

25.4.4 civil commotion, strikes, lockouts – affecting the works and manufacture and transport of necessary goods and materials

25.4.9 the exercise of powers by Government or any statutory power after the date of tender, directly affecting the execution of the works by restricting availability of labour or materials (eg, such as fuel or energy). (Note: this is a new provision)

(ii) *Architect/employer responsibility*

25.4.5.1 compliance with architect's instructions on variations and provisional sums (13), postponement (23), discrepancy between drawings and bills (2.3), nomination of subcontractors (35), suppliers (36) and on antiquities (34)

25.4.5.2 opening up or testing work, etc under clause 8.3 where the contractor is proved not to be at fault.

25.4.6 failure to receive necessary architect's instruction, etc in due time (written applications for instructions to be made neither too early nor too late having regard to the completion date)

25.4.8 where work not forming part of the contract is executed by the employer or a person engaged by him (see clause 29) or the employer supplies materials as agreed (or fails to execute or supply)

25.4.12 any failure by the employer after a requisite notice by the contractor to give access etc to the site over land, etc in his possession or control in accordance with the contract bills or drawings, or as agreed between architect and contractor (Note: this is a new provision) (*Note*: Failure to give possession of the site is not covered. This would constitute a breach of contract.)

(*Note*: Subsections .5, .6, .8, and 12 would give grounds for a claim under clause 26 – loss and/or expense. – See section 3.6.13)

(iii) *Organisational causes*

25.4.7 delays by nominated subcontractors and nominated suppliers (the contractor having taken all practicable steps to avoid or reduce)

25.4.11 in regard to statutory obligations of local authorities or statutory undertakers – the carrying out or failure to carry out work.

25.4.10 the inability of the contractor for reasons beyond his control and which he could not reasonably have foreseen at the date of tender, to obtain labour (25.4.10.1) or materials (25.4.10.2), or both, essential to the proper carrying out of the works. (This is no longer an optional clause).

Note: It is important to note that the restriction of fluctuations (clauses 38.4.8 and 39.5.8) and price adjustment (clause 40.7) during the period where the contractor is in default over completion is conditional upon clause 25 operating unamended and the architect giving all required decisions on the completion date in respect of every written notification by the contractor (see also sections 4.2.5 and 4.5).

3.6.10 PARTIAL POSSESSION

The contract enables the employer before practical completion to take possession of part of the work with the consent of the contractor (clause 18). When this is agreed the following results:

(i) Within seven days the architect certifies for the purpose of this clause the approximate value of the work taken over.
(ii) Practical completion and the defects liability period date from the partial possession. Defects made good to be certified separately.
(iii) Fire, etc risk under clause 22A on the value of work handed over passes to the employer, the contractor's cover to be reduced accordingly.
(iv) The liquidated and ascertained damages figure and retention are pro-rata reduced.

In the 1980 edition the provisions on payment (proportionate retention release on partial possession) are contained in clause 30.4.1.

3.6.11 SECTIONAL COMPLETION

A supplement to the Standard Form with Quantities (Private and Local Authorities editions) was published in December 1975 to adapt the contract to arrangements notified at the tender stage for the works to be completed by phased sections, on practical completion of which the employer takes possession. This is in contrast to the provisions for partial possession referred to above, and provided for in clause 16 (new JCT clause 18) whereby the employer, with the consent of the contractor, may agree at any time before practical completion to take possession of part of the work.

The following are the principles in the 1975 Supplement, which has been updated in line with the 1980 Standard Form.

The use of the adapted contract requires that the tender documents (drawings and bills) identify the sections which comprise the whole works, these sections to be serially numbered and details of each section set out in the Appendix, that is: section number for the purposes of clauses 18 and 22A: the value per bills taking into account the apportionment of preliminaries, etc; defects liability period; date for possession and completion; damages. Work common to several sections should itself be given a separate section. The Appendix must also indicate whether arbitration may commence after practical completion of the whole of the works or of the particular section in dispute.

It must be emphasised that despite phased sections this still remains a single contract with one final certificate. The drawings and bills describe both the work to be done and the divisions into sections for phased completion. There is provision for a certificate of practical completion for each section, the first half of the retention money relating to each section to be released at this stage and a separate defects liability period then to operate. In due course a certificate of practical completion would be issued for the whole of the works from which date the period for the preparation of the final account etc, would run.

An addition to clause 22A makes it clear that on the practical completion of any section the prescribed risks pass to the employer, the contractor to reduce the insurance cover by the value of the section as shown in the Appendix (or other value, agreed and recorded as an amendment to the Appendix).

3.6.12 DEFECTS

Defects, shrinkages or faults appearing within the defects liability period – six months unless otherwise stated – due to materials or workmanship not in accordance with the contract, or to frost occurring before practical completion (or later damage was caused earlier) are the extent of the contractor's liability under the contract for these causes. The period starts on the day named in the certificate of practical completion, the issue of which should be assiduously followed up by the contractor because of its bearing on the defects period, final measurement, retention money and fire insurance cover. Not later than 14 days after the end of the prescribed period the contractor must receive the architect's schedule of defects (clause 17.2). If the delivery of the schedule is late this should be pursued by the contractor who could refuse to do the work, bearing in mind, of course, his common law liability. Once the schedule is delivered, no other items can be added to it (clause 17.3).

Instructions may be issued by the architect to do defects work at any time during the defects period (clause 17.3). This and work in the defects list must be done promptly at the contractor's expense, unless the architect instructs otherwise. Once the architect is satisfied he should issue a certificate of completion of making good defects (clause 17.4). (See clause 30.4 for the treatment of retention money – section 3.8.2 later in this chapter.)

3.6.13 LOSS AND/OR EXPENSE

Payments under the contract may not always fully compensate the contractor for direct loss and/or expense incurred. Clause 26 provides limited scope for a claim where, for the reasons set out, the regular progress of the work is materially affected, thus causing direct loss or expense to the contractor not otherwise recoverable under the contract. This is without prejudice to other rights and remedies the contractor may possess.

The causes of delay affecting the progress of the contract, with loss to the contractor, giving grounds for a claim under clause 26 are as follows:

(i) instructions, drawings, etc properly applied for but not received in time
(ii) where work is required to be tested under clause 8 and is subsequently found to be satisfactory
(iii) there is a discrepancy or divergence between the drawings and the bills (clause 2)
(iv) the employer or the contractors engaged by him on other work have caused delay through the execution or failure to execute work (clause 29)
(v) work is delayed or postponed either at the start or during the progress of the works by instructions from the architect (clause 23)
(vi) agreed requisite access has not been given by the employer
(vii) the architect has under clause 13 issued an instruction requiring a

variation or the expenditure of a provisional sum other than work under clause 13.4.2 (for further information on this item see section 3.6.4 – 'Variations').

This provision emphasises the need for efficient records and contract planning such as progress schedules, preferably agreed with the architect. Requests for instructions, details, drawings, etc must be made in good time on a realistic programme. It should be noted that the programming and organisation of the work is a matter for the contractor to decide unless at tendering stage there are any special conditions for carrying out the work or the employer's requirements set out in the bills are the subject of a variation under clause 13 to be valued accordingly under clause 13.1.2 (see section 3.6.4). A master programme may be required under clause 5.3.

The contractor's remedy if he suffers or is likely to suffer direct loss and/or expense through the regular progress of the work being affected is a prompt written application to the architect stating that he has incurred or is likely to incur direct loss and/or expense under this clause. He should not wait until he can define precisely how progress has been affected. If the architect is of the opinion that regular progress has been affected or is likely to be affected he must determine for himself or through the quantity surveyor the amount he considers should be added to the contract sum, and this should be included in the next interim certificate without the deduction of retention money.

The new provision in clause 26.4 requires nominated subcontractors with claims on this basis (set out in similar terms in NSC/4, clause 13.1) to proceed through the main contractor who will submit applications to the architect for a decision, similar procedures to be followed for ascertaining the claim.

The contractor in making prompt application must specify the grounds for this claim with any supporting information that may be required. He must also submit such details of loss and/or expense as the architect or quantity surveyor reasonably considers necessary to ascertain amounts due. He must, of course, do all he can to mitigate the loss.

If necessary, to ascertain the loss and/or expense, the architect must state in writing the extension of time for each 'relevant event' under clause 25, which gives grounds for such a claim.

3.6.14 DETERMINATION OF CONTRACT

By the employer

If the contractor wholly suspends the work without reasonable cause; fails to proceed regularly and diligently; or to remove defective work or materials as instructed in writing by the architect and the works are materially affected; if he assigns the contract or sublets work without written consent or (in the Local Authorities edition) fails to observe the fair wages clause, then the architect may give notice specifying the default by registered post or recorded delivery. If, unless he disagrees with the notice and refers it to arbitration, the contractor fails within 14 days to remedy the default or subsequently repeats it, (when a further 14 days' notice is not required), the employer may determine the contractor's employment by registered post or recorded delivery within a further 10 days (clause 27.1). This is without prejudice to any other rights or remedies (eg, sueing for breach of contract).

Bankruptcy of a contractor, a composition or arrangement with creditors, winding-up, other than a reconstruction, the appointment of a receiver or manager for debenture holders will all give grounds for automatic determination. If all parties (including the trustee or liquidator, etc) agree, the employment may be reinstated (clause 27.2).

In the Local Authorities edition, corrupt practices also give grounds for determination of all contracts with the contractor (clause 27.3).

When the employment of the contractor ceases, the employer may ask someone else to finish the contract, retaining for his use huts, plant, materials, etc, on site. If not required, the contractor must remove his property promptly when requested in writing by the architect. The contractor must, if required within 14 days of determination, assign agreements for supply of materials, subcontract work, etc, to the employer (clause 27.4) except where there has been bankruptcy or liquidation. This exception also precludes payment by the employer for goods or work already supplied.

The contractor must bear the loss to the employer resulting from this determination (clause 27.4) and any balance eventually due to the contractor is paid after the completion of the works. Any amount owing is a debt payable to the employer.

By the contractor

If the employer does not pay amounts properly due on certificates within 14 days of issue, the contractor may give notice by registered post or recorded delivery stating that if payment is not made within seven days of receipt, notice of determination of his employment will be served.

Determination may also take place if the employer interferes with, or obstructs the issue of any certificate (clause 28.1) and, under clause 28.1 in the private edition, on the bankruptcy, etc of the employer.

If work other than remedying defects is wholly or substantially suspended for the period set out in the Appendix (generally one month or, in the case of damage by fire, three months) for the following reasons, the contractor may determine his employment under the contract forthwith by written notice to the employer or architect:

- (i) *force majeure* (eg, Acts of God, war, epidemics, national emergency, floods, etc)
- (ii) fire damage, etc to new work, but not if due to negligence of contractor or subcontractors
- (iii) civil commotion
- (iv) architect's instructions on variations, postponement, or divergence between drawings bills etc, unless due to negligence or default of contractor
- (v) delay in receipt of instructions, drawings, etc required in writing at a reasonable time having regard to the completion date

(vi) delay on the part of those engaged by employer on work not forming part of the contract under clause 29
(vii) inspections and tests unless the results are unfavourable under clause 8.3.

Note: Within 28 days of fire damage, etc occurring in existing buildings either party may determine under clause 22C if just and equitable.

After determination, the contractor must promptly remove huts, plant, tools and materials which have not been paid for by the employer. The contractor is entitled to payment for work executed, loss or expense as provided for in clauses 26.1 and 34.3, materials ordered and paid for or to be paid for, cost of removal of plant, etc, and direct loss or damage to the contractor, or any nominated subcontractor resulting from determination (clause 28.2). This is without prejudice to other rights and remedies. The employer must inform the contractor of amounts payable which are reasonably attributable to any nominated subcontractor under clause 28.2.2 who must be similarly informed.

3.7 Indemnities and insurances

3.7.1 INJURY TO PERSONS AND PROPERTY

The contractor must indemnify the employer against any liability, resulting from the death of or injury to any person arising from the carrying out of the works unless due to any act or neglect by the employer or any person for whom he is responsible (clause 20.1). This means the onus of proof is on the contractor.

Except in respect of fire and other risks, where borne by the employer, the contractor is fully liable for, and must indemnify the employer against expense, etc arising from injury or damage to property resulting from the carrying out of the works provided it is due to negligence, etc by the contractor or any subcontractor, their servants or agents (clause 20.2). Here the onus of proof is on the employer. Nuclear perils are excluded (clause 21.3).

The distinction should be noted. The personal injury risk is wider than the property risk.

3.7.2 INSURANCES

Persons and property

Irrespective of the indemnity liability for injury to persons and property referred to above, the contractor is required to insure and to see that his subcontractors insure for personal injuries or death arising from the carrying out of the works and not due to any act or neglect of the employer. This means not only employer's liability cover but also special third party cover. The contractor must also insure, and see that his subcontractors insure, for injury or damage to property real or personal where the contractor or subcontractor has been negligent (third party insurance) (clause 21.1.1). Cover should not be terminated until at least the end of the defects liability period.

However, since the normal third party policy does not usually cover all the prescribed risks which include the contract works an 'all risks' policy is also necessary. Cover should be sought where appropriate against liability under the Defective Premises Act 1972. Under clause 21.1.2 policies and premium receipts must be produced by the contractor and subcontractors when required by the employer but a certificate of insurance, for example, would be normally acceptable.

In the event of the contractor or any subcontractor failing to insure as required, the employer may do so under clause 21.1.3 and recover the cost from sums due to the contractor.

Insurance for personal injury to, or the death of, any employee of the contractor or any subcontractor must comply with the Employers' Liability (Compulsory Insurance) Act 1969. For all other insurances under this clause the amount of the cover for any occurrence or series of occurrences arising from any one event should be indicated at the tender stage and inserted in the Appendix to the contract.

Where required under a provisional sum in the bills, claims against the employer for damage to property, other than the works, due to collapse, weakening of support, etc must also be covered by a joint named policy with an insurer approved by the employer (local authorities edition) or architect (private work edition) who retains the policy and premium receipts (clause 21.2). The employer may insure if the contractor fails to do so. Premiums paid by the contractor are added to the contract sum (clause 21.2.3). Cover does not extend to damage due to the negligence of the contractor or any subcontractor (a normal public liability risk), or to design faults, damage reasonably foreseen to be inevitable, to nuclear or war risks, to sonic booms, or to fire, etc risk under clause 22B and C (now known as "Clause 22 perils").

Nuclear perils are an excepted risk in the requirements to insure under clause 21 for damage to works, property, etc (clause 21.3). There is a statutory liability on those engaged in nuclear activities.

Fire etc

There are three clearly defined and separate arrangements for fire, etc under clause 22: (these risks are now described as "clause 22 perils" and fully defined in clause 1.)

(A) a new building where the contractor insures

(B) a new building where the employer bears the fire etc risk and in the private edition is required to insure

(C) alterations, etc to an existing building where the employer is responsible and is required to insure.

Where the contractor is required to insure, it is for the whole of the work, including nominated subcontract work but excluding huts, plant, tools etc. It must be covered by a policy in the joint names of the employer and himself for cases of fire, storm, tempest, flood, riot, etc (but excluding nuclear perils) for the full reinstatement value of the work executed, plus a percentage for professional fees where required in the Appendix, plus unfixed materials on site. If any insurance difficulty arises (eg, extent of cover) this should be sorted out and agreed at the tender stage, the clause to be amended accordingly. Liability to insure ceases at the date of the issue of the certificate of practical completion.

The obligation to insure may be met by an annual or blanket fire policy, with the employer's interest in the policy endorsed thereon, and evidence of this must be produced when required.

Alternatively, policies with a company approved by the employer (architect) together with premium receipts must be deposited with the employer (architect) with the usual provision for the employer insuring in the event of default, the cost to be recovered from the contractor (clause 22A.2).

Note: On determination of the employment of the contractor, withdrawal of cover should be agreed with the employer.

Inflation makes it essential to keep under review 'the full value of the work executed'. Cover must allow for rising costs of building and fees, not only while the work is being carried out but also for increases in the time interval for reinstatement. Any shortfall through inadequate cover would fall on the contractor who, on restoring work damaged, is only entitled to be paid by interim certificates out of insurance monies received (clause 22A.4).

Under clause 22B (private edition) where the employer bears the risk, the contractor may insure if the employer fails to do so and the amount of the premium is added to the contract sum.

In buildings where the employer bears the risk of damage by fire, etc, clearing debris, restoration and replacement of unfixed materials shall be treated as a variation on the instruction of the architect under clause 13.2. *Note*: There is no requirement for the employer to cover by insurance in clause 22B of the local authorities form.

In existing buildings, damage by fire, etc to existing structure and contents, the works and unfixed goods and materials are at the sole risk of the employer. Adequate insurance must be maintained by the employer, but if in the private edition he fails to produce evidence of this insurance, then the contractor may insure the existing structure and contents and the works and unfixed goods and materials in the name of the employer and add the premium cost to the contract sum. Where equitable, within 28 days of loss or damage by fire, etc, the contract may be determined by either party by registered post or recorded delivery. If objected to within seven days, written notice of reference to arbitration may be given. Otherwise the contractor must, if required by the architect, proceed with the work of making good, reinstatement, removing debris, etc, and this work is valued under clause 13.2 as a variation (clause 22C).

Where loss or damage occurs under clauses 22B and 22C, the exact location and nature of the damage must be notified immediately in writing, both to the architect and the employer.

Contractor's plant, etc, is a separate responsibility and does not come within the scope of this clause.

3.8 Certificates and payments

3.8.1 ARCHITECT'S CERTIFICATES

A list of certificates which the architect must issue is summarised below and dealt with elsewhere in the chapter more fully. These certificates must be issued to the employer with an immediate duplicate copy to the contractor (clause 5.8).

(a) Practical completion (clause 17.1): nominated subcontract (clause 35.16) (*Note*: the contract conditions do not precisely define 'practical completion')
(b) Completion of making good defects (clause 17.4)
(c) Approximate value of work on partial possession (clause 18)
(d) Interim certificates – (clause 30.1); final ascertainment of nominated subcontract sums (clause 30.7)
(See also clause 22.A.4 – insurance monies)
(e) Retention money in interim certificates (clause 30.4)
(f) Final certificate (clause 30.8).

The following are among the important matters on which the architect must also certify in writing as and when appropriate:

(i) that the contractor has failed to complete the works by the completion date (clause 24)
(ii) extensions of time by fixing later completion date (clause 25)
(iii) amounts included in certificates under clause 30 for nominated subcontractors which must be notified to them (clause 35.13)
(iv) earlier final payment to nominated subcontractor (clause 35.17)
(v) failure of the contractor to provide reasonable proof that he has paid over to nominated subcontractors amounts included in certificates (clause 35.13)
(vi) when a nominated subcontractor fails to complete the subcontract works on time (clause 35.15)
(vii) loss caused to the employer through determination (clause 27.4.4).

3.8.2 PAYMENTS UNDER THE CONTRACT

Before dealing in detail with clause 30 (payments) of the 1980 edition of the Standard Form it may be of value to look at the inter-relationship between the provisions in all the forms on this rather complex aspect of the conditions of contract. The main provisions on payment are as follows:

(1) *Main contractor*
The terms of clause 30 of the Standard Form reviewed below only cover payments by the employer to the main contractor even though they will include amounts due to nominated subcontractors which the contractor must pass on in accordance with the requirements of the subcontract.

(2) *Nominated subcontractors*

(a) Clause 35.13, 17, 18 and 19 of the main contract deals with payments to subcontractors by the main contractor – amounts due, retention, early final payment and direct payment by employer (see section 6.3.3).

(b) Clause 21 of NSC/4 (or 4a) deals fully with the payment of nominated subcontractors under the 1980 edition of the Standard Form including amounts due, retention, payment, right to suspend work (section 6.4.6).

(c) Under NSC/2 (or 2a where appropriate) clause 5.1, the employer undertakes that the architect will operate the early payment provisions of clause 35.17 to 19 of the main contract and clause 35.13 on direct payment on default of the contractor (section 6.2.1).

(3) *Domestic subcontractors*

The terms of payment are entirely a matter between the contractor and any subcontractor to whom work is sublet under clause 19. Normally the conditions in clause 13 of the non-nominated (blue) subcontract form would apply (section 5.4.3).

In revising clause 30 (certificates and payments) in the 1980 edition, the Joint Contracts Tribunal has taken the opportunity of setting out in detail what has to be included in interim certificates and how the contract sum is to be adjusted to enable the final certificate to be issued. The changed procedures for retention are explicit and include financial statements, employer's right of recourse and, in the private edition, the opening of a separate bank account if requested. To give a complete picture of the arrangements for payment, this section should be considered in conjunction with the provisions in respect of nominated subcontractors referred to above.

Interim certificates

To enable the required interim certificates to be prepared the architect may arrange for an interim valuation by the quantity surveyor whenever he considers it necessary (clause 30.1). (For formula adjustment this is essential – see clause 40.2.1.) The Appendix sets out the period for the issue of interim certificates (normally monthly) up to the issue of the certificate of practical completion. Thereafter and not more frequently than monthly, further interim certificates are to be issued as and when further sums are due to the contractor. Payment is to be made within 14 days from the date of issue (clause 30.1 – see also clause 5.8). Interim certificates are not conclusive evidence that work materials, etc are in accordance with the contract (clause 30.10).

Subject to any agreement on stage payments, the amounts due up to seven days before the date of any interim certificates are as follows:

Clause 30.2.1 — retention deducted

1 Total value of work properly executed by the contractor including

variations under clause 13.2 and formula adjustment under clause 40.

2 Materials on or adjacent to site for incorporation by main contractor delivered reasonably, properly and not prematurely and adequately protected.

3 Off-site materials in accordance with clause 30.3, where the architect exercises his discretion and all conditions have been met (ie, materials ready and intended for incorporation, separated, identified and fully marked, no restriction in title, fully insured for clause 22 risks until delivered to site – see section 3.6.3).

4 Amounts due to each nominated subcontractor under clause 21.4.1 of NSC/4 or 4a.

5 Contractor's profit at bill (or relevant) rates on payments to nominated subcontractors under clauses 30.2.1.4 and 30.2.2.5 less deductions under clause 30.2.3.3.

Clause 30.2.2 — no retention deducted

1 Amounts due as a result of costs incurred by the contractor under clauses 6, 7, 8, 9, 17 and 21.2.3 (see final certificate computation (.11) below for details).

2 Loss and/or expense claims ascertained under clauses 26.1 or 34.3.

3 Early final payment to a nominated subcontractor under clause 35.17.

4 Fluctuations payable under clauses 38 or 39.

5 Amounts due to each nominated subcontractor under clause 21.4.2 of NSC/4 or 4a.

From the total of these items shall be deducted under clause 30.2.3:

1 Sums due to the employer under clause 35.18.1 (where defects appear in the nominated subcontract work before issue of the final certificate – see section 6.3.7) and clauses 22A.2 and 35.24.6.

2 Any amounts allowable to the employer under the fluctuations provisions of clauses 38 or 39.

3 Amounts in respect of a nominated subcontractor under clause 21.4.3 of NSC/4 or 4a.

From these are deducted retention under clause 30.4 and amounts certified in earlier interim certificates.

Note: Clause 21 of NSC/4 or 4a referred to above is in similar terms to clause 30.2 of the main contract and is dealt with in section 6.4.6.

Clause 3 makes it clear that any amount which under these contract conditions is to be added to, deducted from or adjusted in the contract sum must be included in the next interim certificate after the amounts have been ascertained.

Retention

The retention percentage is 5 per cent unless a lower rate has been agreed and specified in the Appendix (eg not more than 3 per cent is recommended where the contract sum is £500,000 or over) (clause 30.4.1). This percentage may be deducted from all sums ascertained under clause 30.2.1 (ie, including amounts due to nominated subcontractors) but not from sums under clause 30.2.2 (see above).

The rules for deduction and release are as follow:

Deduction

(a) Full percentage (5 per cent or lower if agreed) deducted from amounts relating to work which has not reached practical completion (see clauses 17.1, 18.1.2 (partial possession) or 35.16 (nominated subcontractors)) including the value of on and off-site materials under clauses 30.2.1.2, 3 and 4.

(b) Half percentage deducted from amounts relating to work which has reached practical completion but in respect of which a certificate of completion of making good defects has not been issued, nor an interim certificate issued for early final payment to a nominated subcontractor.

(c) No retention is deducted from any subsequent payment.

Amounts retained would be reduced by any release on partial possession under clause 18 or on early final payments to nominated subcontractors (clause 35.17).

Release

The effect of these provisions is that, subject to the employer's right of recourse against retention explained below, the contractor will have released to him one half of the retention money in the next interim certificate after practical completion. The second half will be released in the next interim certificate after the end of the defects liability period or the issue of the certificate of completion of making good defects whichever is the later. These provisions differ slightly from the pre-1980 edition where retention was released on certificates specifically issued at the same time as the certificates of practical completion and of making good defects.

Statements must be prepared by the architect or quantity surveyor at the date of each interim certificate setting out the retention in respect of the main contractor and each nominated subcontractor in arriving at the amount of the certificate. Copies of statements are to be sent by the architect to all parties concerned including the employer. (clause 30.5.2)

Employer's interest

The employer's interest in the retention is fiduciary as trustee for the contractor and any nominated sub-contractor, without any obligation to invest (clause 30.5.1). However in the private edition of the Standard Form the employer must place the amount deducted in a separate designated bank account if the contractor or any nominated subcontractor so requests at the date of payment under each interim certificate, and he must certify to the architect, with a copy to the contractor, that this has been done. Only the employer is entitled to any interest accruing and he is not required to account for it (clause 30.5.3). There would appear to be nothing to prevent a contractor requesting the use of this provision on the first certificate and to apply to all future certificates.

Note: The main contractor, in turn, has a fiduciary interest in retention monies due to nominated subcontractors. This is set out in clause 21.9 of NSC/4 (and 4a) – see section 6.4.6.

Employer's right of recourse

In respect of certain costs recoverable from the contractor, the employer has a right of recourse against amounts due. For example, the following clauses state that for the reasons set out the employer may deduct amounts referred to from any monies due or to become due to the contractor under this contract:

4.1	contractor's failure to comply with architect's instructions (eg, on rectifying defects)
21.1	insurance of persons and property
22.A.2	insurance against fire etc
24.2	liquidated and ascertained damages
27.4	determination of contractor's employment.

There are also obligations to deduct amounts under clause 31 (tax deduction scheme) and provisions for monetary adjustments under the VAT Agreement. This is separate and distinct from deductions in computing the amount of interim or final certificates under clauses 30.2.3 and 30.6.2.

The ways in which deductions can be made from amounts due are set out in clause 30.

(a) *Interim and final certificates*
Under clause 30.1.1.3 the contractor must be informed in writing of the reason for any deduction. Where this right is exercised against retention see (b) below

(b) *Retention money*
Notwithstanding the fiduciary interest of the employer in the retention referred to above, under clause 30.1.1.2 the right to deduct from amounts due or to become due by inclusion in interim certificates under clause 30.4 extends to retention money. The restriction in clause 35.13.5.4.2 applies (ie, in any direct payments to nominated subcontractors under clause 35.13.5.4.3, the amount available shall be limited to retention due to the contractor – see section 6.3.3).

The right to deduct is governed by the terms of clause 30.5.4 under which the contractor must be informed of the amounts deducted from the retention of the main contractor or any nominated subcontractor (by reference to the latest statement required with each interim certificate under clause 30.5.2.1). Where a subcontractor's default has given rise to a deduction, to that extent the contractor may pass it on to the subcontractor (see NSC/4 clause 21 – section 6.4.6).

Final certificate

To determine the final amount due under the contract, leading to the issue of the final certificate, the contractor must send to the architect (or the quantity surveyor if so instructed) all necessary documents including accounts of nominated subcontractors and suppliers either before, or within a reasonable time after, practical completion (clause 30.6.1).

Within the period of final measurement and valuation stated in the Appendix (normally six months from practical completion) a statement of the final valuation of all variations must be prepared by the quantity surveyor, including those relating to nominated subcontract work, a copy to be sent to the contractor and relevant extracts to each nominated subcontractor.

The computation of the adjusted contract sum on the basis set out below must be sent to the contractor before the issue of the final certificate (clause 30.6.3).

The contract sum must be adjusted as follows under clause 30.6.2, the numbers given being subparagraph references:

Deductions

1. all prime cost sums and amounts in respect of named subcontractors under clause 35.1 and contractor's profit thereon as stated in the bills
2. amounts due to the employer under clauses 35.18.1 (rectification of defects in nominated subcontract work), 22A.2, 35.24.6.
3. provisional sums and work described as provisional in the bills
4. valuation under clause 13.5.2 of omissions, the subject of an architect's instruction under clause 13.2 (see also reference to clause 13.5.5).
5. any fluctuations *allowable* to the employer under clauses 38, 39 or 40.
6. any other amount required to be deducted under the contract.

Additions

7. total amounts included in interim certificates for final payments to nominated subcontractors under clause 30.7 (see below) in accordance with the terms of NSC/4 or 4a
8. value of work by main contractor tendered for and accepted under clause 35.2 (work otherwise reserved for a nominated subcontractor)
9. nominated suppliers' accounts (including 5 per cent cash discount but excluding VAT treated as input tax by the contractor) (clause 36)
10. profits at billed or other agreed rates on amounts paid to nominated subcontractors and suppliers and for work under clause 35.2 (see 7, 8 and 9 above)
11. amounts payable to the contractor as a result of costs incurred under the following clauses:

    6.2 local authority, etc fees and charges not priced in bills

    7 setting out, etc

8.3 tests of work; materials – not faulty
9.2 royalties, etc following architect's instruction
17.2.3 defects – architect's instruction – cost added to contract sum
21.2.3 provisional sums – joint named insurance cover

12. value of variations under clause 13.5 (other than omissions)
13. valuation of contractor's work or disbursements, on the expenditure of provisional sums included in the bills or work described as provisional
14. loss and/or expense claims ascertained under clauses 26.1 or 34.3
15. amounts under the private edition payable to the contractor under clauses 22B or 22C
16. fluctuations payable under clauses 38, 39 or 40.
17. any other amount required to be added under the contract.

This computation relates to matters affecting the final ascertainment of the contract sum. It does not cover liquidated damages or other deductions from the contract sum where the contractor is in default (see above – 'Employers' right of recourse').

Issue of final certificates
As soon as practicable, and at least 28 days before the final certificate is issued, an interim certificate must be issued under clause 30.7 (even if within one month of the previous interim certificate) clearing all final amounts due to nominated subcontractors under NSC/4 or 4a and still outstanding, less any credits under NSC/2 and 2a clause 2.2. The final certificate will therefore only deal with monies due to or from the contractor. From it can be deducted direct payments to the nominated subcontractor of final amounts certified under clause 30.7 and not paid over. (*Note*: Clause 35.17 provides, in addition, for early final payment – see section 6.3.7.)

As soon as practicable and before the expiry of the period (normally three months) stated in the Appendix from the end of the defects liability period, the completion of making good defects under clause 17, or the supply by the contractor of all necessary documents for the final account, whichever is the latest, the architect must issue the final certificate and inform each nominated subcontractor of the date of issue (clause 30.8). The final certificate must state:

(a) the total of the amounts already stated as due under interim certificates
(b) contract sum adjusted as above
(c) difference being the balance due, being the debt payable to or by the contractor as from the fourteenth day after the date of the certificate.

(This would not preclude recovery on an outstanding interim certificate.) There is thus a clear timetable for the settlement of contracts but this calls for efficiency and business-like methods on the part of architect, quantity surveyor and contractor.

Conclusive nature
Under clause 30.9, except in the case of fraud, the final certificate is conclusive

evidence in arbitration, or any proceedings arising out of the contract, that:

(i) The quality of materials and the standard of workmanship are to the reasonable satisfaction of the architect wherever required (see clause 2.1). (Latent defects due to the work not complying with the contract will be the contractor's responsibility).

(ii) The contract sum is as finally certified except for, and only to the extent of, any error in arithmetic or the accidental inclusion or exclusion of work, materials or of figures in any computation.

The limited nature of this provision should be borne in mind when considering the remedies of the employer.

The conclusive nature of the final certificate is subject to certain provisos where arbitration or other proceedings have been commenced either before or within 14 days after the issue of the final certificate as follows:

(a) *Before its issue*: The final certificate will become conclusive subject to the earlier of the following: the terms of any award etc., on the conclusion of proceedings; or terms agreed in a partial settlement if at the end of 12 months the parties have not proceeded with the action.

(b) *Within 14 days of its issue*: The final certificate is conclusive except for points the subject of proceedings.

Delays in honouring certificates

A frequent problem, particularly at a time when credit is scarce and expensive, is the failure to meet the payment requirements under the contract. If there is delay in honouring certificates, the employer should be reminded of his contract obligations and the assistance of the architect sought. There is no right to suspend work as there is in subcontracts (NSC/4 and 4a clause 21.8 (section 6.4.6): see also 'green' form clause 11(e) and 'blue' form clause 13 (section 5.4.3))

Serious and persistent delays in the issue of certificates or under certification of work could give grounds for an immediate reference to arbitration under Article 5.2.2 and, if the employer interferes in the issue of certificates, for determination under clause 28.1.2. There is no contractual provision for interest on overdue payments but it could be claimed that failure to pay on time was a breach of contract and the interest sought was a measure of the damages suffered.

3.8.3 NOTE ON LIMITATION ACTS

The conclusive nature of the final certificate is so limited that a note on the Limitation Acts could usefully be read in conjunction with the provisions of clause 30.9.

Common law remedies for breach of contract must be pursued within six years from the cause of the action if the contract is under hand, and twelve years if it is under seal, after which action is statute barred.

Following the case of *Anns V Merton London Borough Council* the position would appear to be that in an action for negligence the period runs from the default becoming apparent and damage resulting. This means that years may pass before the cause of action appears, raising difficult questions of proof and the need to keep records almost indefinitely, as well as the effect of the lapse of time on costs and the continued existence of the defendant (eg, if the contractor has gone into liquidation). (While professional indemnity insurance may be available for defective design, there is virtually no cover for defective workmanship. Third party cover for damage to persons or property resulting from defects, if proved, could be covered by public liability insurance.)

The contrasting legal position under the Defective Premises Act 1972 should be noted (see section 7.5.1).

Another facet of this problem is the practice of executing main contracts under seal and subcontracts under hand. Since the limitation period would be shorter for subcontract work, to the detriment of the main contractor in any recourse against his subcontractor, it is often advised that main and subcontract documents should be similarly executed (ie, both under seal or under hand). It has been argued, however, that by reason of the terms of clause 3 of the 'green' form on indemnities to the main contractor, for example, the sub-contractor would not escape liability if the limitation period were longer under the main contract.

3.8.4 VALUE ADDED TAX

VAT is dealt with by means of clause 15 and supplemental provisions (the VAT agreement) which are incorporated in the contract, and does not require execution separately.

Clause 15 makes it clear that the contract sum is exclusive of any VAT and the contractor is entitled to recover from the employer the equivalent of any loss of input tax if the supply of goods and services to the employer subsequently becomes exempt after the date of tender.

New work is zero rated, other work is taxable at the current rate. This does not add to the cost of the work insofar as the contractor is only the agent of Customs and Excise, collecting and accounting for this tax. The employer agrees to pay the appropriate tax to the contractor (Supplemental provisions SP1). At the time of each certificate the contractor gives a written provisional assessment of the value of the work, less retention, which is zero rated and taxable, with a note of the rate and the grounds on which it is chargeable (SPI). (This is not a tax invoice.) The employer calculates the VAT due at the date of payment and includes it in his cheque honouring the certificate. The contractor then issues a tax receipt (SPI.4). Further payments may be suspended if a validating receipt required by the employer is not issued by the contractor (SP7).

As soon as possible after the issue of the certificate of making good defects, a final statement of values, taxable and zero rated, with a note of tax already received, must be prepared by the contractor (SP1.3) liquidated and ascertained damages being disregarded in the calculations (SP2). The employer then calculates the balance due and pays within 28 days – or, if overpaid, secures a refund from the contractor (SP1.3).

If the employer disputes any provisional assessment and notifies the contractor in writing within three working days the contractor must, within another three days, reply in writing either withdrawing or confirming the assessment.

If the employer does not agree with the final statement he may, before payment, ask the contractor to refer the matter to the Commissioners of Customs and Excise for a decision. If he is dissatisfied with this decision the contractor, on being indemnified by the employer, must appeal. Settlement must be effected within 28 days of the results of this appeal (SP.3). The arbitration provisions in Article 5 do not apply to disputes under clause 3 of the agreement (SP6). (*See* JCT notes on this agreement and Practice Note 17.)

3.8.5 CONSTRUCTION INDUSTRY – TAX DEDUCTION SCHEME

Outline of scheme

The current Inland Revenue scheme for deduction of tax from payments to subcontractors came into effect on 6 April 1977 and includes payments on and after that date under contracts then current.

Provision for this scheme was made in the Finance (No 2) Act 1975 (sections 68 to 71 and schedules 12 and 13) and in the Income Tax (Subcontractors in the Construction Industry) Regulations 1975 (SI 1975: 1960). An Inland Revenue booklet is available entitled *Construction Industry: Tax Deduction Scheme*.

The scheme is far-reaching and differs from the 1972 scheme in that it includes limited companies and applies to all subcontracts (other than for materials only). The definition of 'subcontractor' is very wide since it extends to main contractors working for 'contractors' (ie, public bodies and private clients with their own permanent building departments, etc).

Official publications and the legislation should be consulted for authoritative information but some salient features of the scheme are:

(i) Unless the 'contractor' is satisfied that the 'subcontractor' has a current valid tax certificate, all payments on and after 6 April 1977 made under a contract for construction operations must be subject to a tax deduction at the standard rate (the direct cost of materials as defined in the scheme, which are to be notified by the subcontractor or, failing that, estimated by the contractor, are not subject to tax).

(ii) Certificates should be applied for by all firms in the construction industry (both building and civil engineering). These certificates are identified as 714 I, 714 P, and 714 C (for companies). The date of expiry is shown on the certificate. Contractors should be immediately informed by the subcontractor when he no longer holds a valid certificate.

(iii) As indicated above the definitions of 'contractor' and 'subcontractor' under the scheme are much wider than those normally understood in the industry.

(iv) Administrative arrangements are extensive:

(a) Tax deducted (or which should have been deducted) for which the contractor has to give the subcontractor an SC 60 receipt must be accounted for monthly along with PAYE tax.

(b) Payments will be made gross where the contractor is satisfied that the subcontractor has a current valid tax certificate. Subcontractors holding I (individual) or P (partnership) certificates must issue 715 receipts within seven days. For companies (with a 714C) there is an alternative 'certifying document method', payments being made into a nominated bank account.

(c) Annual returns must be made by contractors and subcontractors. If tax deductions from subcontractors exceed tax liability the balance will be refunded.

Contract provisions
Although the scheme is legally binding and requires no contractual provision the JCT considered it desirable to amend the Standard Form to ensure the smooth running of the contract where the scheme operates (ie, the main contractor is a 'subcontractor' because the employer is a public body, etc). There is no comprehensive timetable for the requirements of the scheme but one is set out under the relevant contract provisions together with certain remedies and protection for the parties not otherwise available under the legislation. Clause 30B was introduced in November 1976, Practice Note 22 being issued at the same time to give a brief description of the scheme and explain the purpose of the amendment. The JCT has prepared identical clauses for its Fixed Fee, Design and Build and the Minor Works Forms (see Chapter 7). Similar amendments were introduced for the subcontract form – for the nominated ('green') form clauses 10C and 10D; for the non-nominated ('blue') form clauses 12C and 12D (see Chapter 5). In the 1980 JCT nominated subcontract forms the clauses are 20A and 20B of NSC/4 and 4a (see Chapter 6).

In the 1980 edition of the Standard Form the clause number is now 31. It starts off with some definitions in subclause 1. The following subclause 2 refers to the Appendix to the contract which now sets out whether the employer was or was not a 'contractor' under the scheme at the date of tender (when of course the main contractor becomes a 'subcontractor'). It is then stipulated that if the employer is not a 'contractor' the following subclauses 31.3 to 31.9 do not apply. If at a later date the employer does become a 'contractor' the main contractor is to be informed at once, subclauses 31.3 to 31.9 then to operate. By the reproduction of the Appendix in the preliminaries in the bills of quantities the contractor will be clearly informed at the date of tendering on the status of the employer under the scheme, Recital 4 referring to the Appendix on this point.

Subclause 31.3 to 31.9 provide as follows:

31.3 Not later than 21 days before the first payment is due, the main contractor must provide satisfactory evidence that he is entitled to be paid gross or alternatively inform the employer in writing (with a copy to the architect) that he is not. If he is not satisfied with the evidence for payment gross the employer must in writing notify the contractor within 14 days of submission of evidence that payments will be net giving reasons for the decision. (See clause 31.6 which then must be observed.)

31.4 If at a subsequent date a valid certificate is obtained the employer must be informed at once clause 31.3.1.1 then to apply. Where a certificate expires before final payment, then the contractor must give the employer evidence not later than 28 days before expiry that from the date of expiry payments may still be made gross. Payments must be made net where the employer is not satisfied or the contractor has informed the employer in writing that future payments should be made net (ie, there is no longer a valid tax certificate). If a certificate is cancelled the employer must be informed at once with a note of the date.

31.5 To protect the contractor the employer must promptly send to the Inland Revenue 715 vouchers given by contractors (holding I or P certificates).

31.6 Where payments are to be made net, the employer must notify the contractor in writing, and require him to state not later than seven days before each future payment becomes due the amount in the payment which represents the direct cost of materials (which is not subject to tax). (Since this would include payments due to subcontractors they are required to provide similar information under the terms of their subcontract.) Where an incorrect statement is made the employer must be indemnified against loss or expense (eg, because of inadequate deduction). If no statement is supplied by the contractor the employer is entitled to make a fair estimate.

31.7 Errors or omissions must be corrected by the employer, subject to any statutory obligations not to do so.

31.8 These provisions (being statutorily enforceable) prevail over any other conditions in the contract.

31.9 The arbitration provisions under Article 5 apply to any dispute, etc unless there is any statutory alternative laid down under tax legislation.

If failure to comply with these conditions affects payments (eg, delay in supplying information) the responsibility would be the contractor's. Practice Note 22 gives guidance, approved by the Inland Revenue, where direct payment is made to a nominated subcontractor by the employer under the terms of the contract:

(i) *Where the employer is not a 'contractor'*
The Scheme does not apply and payment is to be made gross. Details of

payments may be required by the Inland Revenue at the end of the year.

(ii) *Where the employer is a 'contractor'*
Payment subject to scheme's provisions to be operated in same way as payments to the main contractor (ie, a 'subcontractor' under the scheme).
(*Note*: The Property Services Agency is not a 'contractor' under the scheme – see Chapter 9.)

3.9 Other conditions

3.9.1 PERSON-IN-CHARGE

A competent person-in-charge (previously 'foreman') must be constantly present in charge of the job, instructions to him from the architect or directions by the clerk of works being regarded as given to the contractor (clause 10).

3.9.2 RATES

Rates on temporary premises such as offices, huts, sheds, etc can be added to the contract sum, unless covered by a bill item, a provisional sum, or a nomination (clause 6.2).

3.9.3 HOSTILITIES

Subject to certain conditions, the contract can be determined on the outbreak of hostilities but the architect may, within 14 days of a notice, issue instructions for protective work or for continuation up to a certain point (clause 32).

3.9.4 WAR DAMAGE

The architect may issue instructions following war damage, the work to be treated as a variation under clause 13.2 (clause 33).

3.9.5 ANTIQUITIES

Fossils, antiquities, etc become the property of the employer. Direct loss and/or expense on cessation of work, extraction, removal, preservation, etc are by addition to the contract sum, reimbursed to the contractor (without retention deduction) where not otherwise recoverable under the contract, and where necessary the architect must state in writing the extension of time under clause 25 relevant to a claim under this clause (34).

3.9.6 ROYALTIES AND PATENT RIGHTS

Under clause 9 the contractor must include in the contract sum any payments for royalties, etc arising from the use of any patented items as set out in the bills and must indemnify the employer against any claim for infringement. Where the use arises from an architect's instruction the contractor is not liable for any infringement, and any royalties or damages payable are added to the contract sum.

3.9.7 PERSONS EMPLOYED BY EMPLOYER
Where the employer, either himself or by the employment of others, wishes to carry out work concurrently with, but not forming part of the contract (ie, not as a subcontract), the contractor must permit this, provided he can assess the effect on his work from the information given in the contract documents. If there is no information in the contract documents, or this is inadequate, the contractor's consent must be obtained by the employer on fair and reasonable agreed terms (this consent should not be withheld unreasonably) (clause 29). The employer would be responsible for this work and for persons employed (eg, injury to persons and property) and resultant delays to the contract works could give rise to a claim for extension of time (clause 25.4.8), loss and/or expense (clause 26.2.4) or even the right to determine (clause 28.1.3.6).

3.10 Arbitration

There is a provision in the Standard Form of Building Contract enabling matters in dispute to be referred to arbitration under the Arbitration Acts 1950 to 1979which, unless otherwise provided, will apply to all proceedings irrespective of the nationality of the parties or the location of the works. In the pre-1980 edition arbitration was covered by clause 35 but in the 1980 edition it is included in the Articles of Agreement. The arbitrator's authority is irrevocable and his decisions, which are final and binding, are reached after observing established legal procedures and principles.

The arbitration provisions in the contract are as follows (references are to Article 5 of the Articles of Agreement).

With the exception of disputes under clause 3 of the VAT agreement or as prescribed under clauses 19A (fair wages – Local Authorities edition) or clause 31 (tax deduction scheme) any dispute or difference over the meaning of the contract, or over matters arising under the contract, which occurs between the employer (or the architect on his behalf) and the contractor, either during progress or after the completion or abandonment of the works, may be referred by either party to an arbitrator (article 5.1). This includes matters left to the discretion of the architect, the withholding of any certificate, adjustments of the contract sum under clause 30.6.2; rights under clauses 27, 28, 32 or 33; or the unreasonable withholding of required consent by any party.

If the parties are unable to agree on an arbitrator within 14 days of written notice of reference to arbitration the President or Vice-President of the RIBA can be asked in writing by either side to appoint (article 5.1). Under article 5.2, unless both parties agree in writing these proceedings do not take place until practical completion, termination of the contractor's employment or abandonment of the works except on:

(a) dispute over the power to issue instructions (clause 4.2)
(b) a certificate being improperly withheld or not being in accordance with the conditions

(a) dispute over the power to issue instructions (clause 4.2)
(b) a certificate being improperly withheld or not being in accordance with the conditions
(c) disputes concerning outbreak of hostilities or war damage
(d) on article 3 or 4 (note: the Private Work form only, provides for a contractor's objection to an appointment following the death of the architect or quantity surveyor)
(e) a dispute or difference over an extension of time
(f) a dispute under clause 4.1 on whether a contractor's objection to a variation under clause 13.1.2 is reasonable.

The effect of arbitration on the conclusive nature of the final certificate should be noted (see section 3.8.2).

The arbitrator's powers are very wide. He may, for example, review and revise certificates and valuations and may direct measurements and valuations, and he may also disregard any opinions, decisions or notices already given.

Fluctuations payments determined under clause 38.4.3 and 39.5.3 and formula adjustments under clause 40.5 would not appear to be subject to arbitration.

As provided in the Arbitration Act 1950 the award is final and binding on the parties (article 5.4).

There is an additional optional provision in the 1980 edition (article 5.1) that if a dispute or difference is substantially the same as one which has been referred to arbitration under the Agreements NSC/2 or NSC/2a (see Chapter 6), or under agreement NSC/4 or 4a between the main contractor and any nominated subcontractor, or under the standard form of tender for nominated suppliers (see Chapter 7), or where sale conditions include clause 36.4.8, the employer and contractor agree to have the matter referred to, and be bound by, any decision of that arbitrator unless either party reasonably considers he is not suitably qualified to decide the issue. Since this provision is optional it must be indicated at tender stage and in the Appendix whether or not the joinder provision will apply.

A provision such as this, optionally allowing joinder in disputes where substantially the same issues are involved, appears in all the arbitration provisions in the following JCT editions, the main contract choice to apply throughout:

Standard Form of Building Contract (referred to here)
Nominated Subcontract Form NSC/4 and 4a (see section 6.4.2)
Employer/Nominated Subcontractor Agreement NSC/2 and 2a (see section 6.2.1)
Nominated Suppliers Tender Form (see section 7.4)

This follows the wording and intent of clause 24 (arbitration) of the pre-1980 nominated ('green') form and clause 27 of the 'blue' form (see section 5.9.3). It avoids a possible risk of two separate arbitrations. However, it will be appreciated that these provisions cannot extend to those against whom action may be taken but who are not parties to these contracts (eg, a material manufacturer). There should be borne in mind the provisions of the Civil Liability (Contribution) Act 1978 which very briefly provides as follows.

(i) From the 1 January, 1980 the entitlement to a contribution from other parties liable for the same damage is extended to a breach of contract where the damage has been suffered on and after that date.
(ii) It applies to court action only and not to awards under an arbitration agreed between the parties.
(iii) It would, however, include a bona fide out-of-court settlement by one of the parties who could then claim a contribution from another party.

On the reference to arbitration awards it should be pointed out that there is no requirement in the contract that arbitration procedure must first be exhausted before reference can be made to the courts, but if action is initiated by one party, the other could apply at once for a stay of proceedings on the grounds that there is an agreement to refer the matter to arbitration. Such an application is likely to be granted. There is no general right of appeal from an award of an arbitrator and the court is unlikely to intervene unless the arbitrator has misconducted the proceedings or made a serious error on a point of law. A prompt application to the court is necessary if either party seeks to have an award set aside. An arbitrator may on a point of law be requested by either party to state a special case for decision by the court.

An interim award may be made before a final award and may be sought where proceedings are likely to be protracted and payment is involved. An award is enforceable as if it were a judgement of the court. The arbitrator directs as to who is to pay the costs.

If a party fails to appear or supply the necessary documents and evidence, the arbitrator, after every effort to scure co-operation, may proceed *ex parte* (ie, in the absence of the party concerned).

Note on Arbitration Act 1979

This act, which received Royal Assent in April 1979, is primarily concerned with procedural aspects of arbitration. It is particularly directed at replacing the power to require an arbitrator to state a special case under section 21 of the Arbitration Act 1950 which could cause serious delays, with a limited right of appeal to the High Court on a point of law. The High Court is enabled to require arbitrators to publish the reasons for their awards in certain circumstances. There is provision for a limited right of appeal from a High Court decision and for agreement made after the arbitration has commenced to exclude the right to go to the court on a preliminary point of law or on appeal.

3.11 Standard form – bills of approximate quantities

The JCT considered that since there was apparently a considerable volume of work by value let on the basis of approximate quantities, the Standard Form should be adapted to meet this situation. In October 1975 the Standard Form of Building Contract for use with Approximate Quantities made its appearance in both private and local authorities editions. This was accompanied by Practice Note 20 explaining the purpose of the new form, drawing attention to major points and listing the differences from the Standard Form. The 1980 editions of the Standard Forms included both Local Authorities and Private Editions of the form with approximate quantities.

Under this procedure tenders are invited on bills which are clearly indicated as only approximate thus necessitating a complete remeasurement with appropriate contract provisions. The Standard Form with Quantities was not suitable because it envisaged adjustments only for such items as variations resulting from architect's instructions and the expenditure of provisional sums. If some sections of the bills are provisional they must be so marked, the Standard Form to be used with remeasurement for these sections.

The total of the prices submtted on the approximate bills is know as the 'tender price' and the final amount due as the 'ascertained final sum'. Effect is given to these terms in the adapted form. The major alterations necessitated by complete remeasurement are considered below in general terms.

3.11.1 BILLS OF QUANTITIES

Since bills are approximate, the recitals state that they describe the work to be done and are intended to give a reasonably accurate forecast of the quantity of the work. This means the work must be substantially designed although not completely detailed; only the bills are approximate. If the quantities are not a reasonably accurate forecast, this raises a number of points. Firstly the rates may not provide a reasonable basis for valuation; in these circumstances the contractor is entitled to a 'fair valuation'. An extension of time would be allowed for undr this heading.

In the definition of 'variation' the reference to a change in the quantity of work is omitted, since this would be reflected in the remeasurement. Provisional and prime cost sums are dealt with in the usual way. In 'Measurement and valuation of work' it is stipulated that all work must be measured and valued, the rules for valuation covering the valuation of all such work measured by the quantity surveyor.

3.11.2 ASCERTAINED FINAL SUM

In calculating the 'ascertained final sum' all documents, etc must be sent by the contractor to the architect or quantity surveyor before practical completion or within a reasonable time thereafter. Priced bills of remeasurement must be supplied by the quantity surveyor to the contractor not later than the end of the period of final measurement and valuation (normally six months). The ascertained final sum is built up by adding to the work as valued in the bills of remeasurement, sums due in respect of nominated subcontractors and suppliers, statutory fees, joint named cover insurances, preliminaries priced in the bills and adjusted where appropriate to the remeasured value of work done, loss and/or expense claims and fluctuations, etc.

A firm price (ie, under clause 38) is not appropriate on approximate bills because of lack of precision for ordering, etc. If clause 40 applies, work groups could not be used because tender weightings on approximate bills would not be accurate.

Interim valuations are always necessary and are therefore not at the discretion of the architect.

4

FIRM PRICE AND VARIABLE PRICE CONTRACTS

Firm price and variable price contracts have so many facets that they require separate consideration.

A firm price contract is one in which the original contract sum is not adjustable for any increases in the cost of labour, materials, etc during the currency of the contract.

A variable price contract makes provision for an adjustment upwards or downwards in the contract sum for stipulated movements in wage rates and other labour costs, materials prices, etc, as defined in the contract.

There are extensive legal and economic considerations and a look at the past helps to explain some of the more important factors.

4.1. Recent history

Early in 1940 a fluctuations clause 25 (a) was introduced into the Standard Form of Contract to meet the uncertain conditions of wartime building. In the unsettled post-war period, this continued almost universally until 1957 when the Government decided to revert to the pre-war practice of firm price tendering, advising local authorities and nationalised industries to do the same. This decision was subject to the two conditions:

(i) that the estimated contract period should be not more than two years
(ii) that the work should be thoroughly planned in advance

The Minister of Works emphasised that variations must be kept to a minimum and tenders accepted quickly. The NFBTE in a statement dated 11 July 1957 set out what it understood by 'thoroughly planned in advance' (*see* Chapter 2), and made it clear that tenders should be accepted within two months, after which they should be subject to confirmation or adjustment by the contractor.

This has become a most controversial subject. Periods of economic difficulty, which made firm prices more hazardous, increased the resentment of contractors, particularly when the initial tendering conditions were ignored. Government departments, however, continued to regard firm price tendering as a bulwark against inflation in spite of criticism. When work was short serious risks were taken by contractors. Bitter experience and a greater volume of work led tenderers to make more adequate provision for imponderables, particularly when prices were rising. The two systems - firm price and fluctuations, were widely used, although almost invariably public authority contracts of up to two years' duration were on a firm price basis. On occasion tenders were invited on both bases so that comparisons could be made before acceptance.

In 1963 the Standard Form of Contract was revised and the introduction of other clauses resulted in the fluctuations clause being renumbered 31 although it differed little in principle from clause 25(a) in the previous edition. In the following three years, inflation increasingly affected long term commitments, and in the 1966 budget, Selective Employment Tax was introduced. This caused serious concern. Fiscal decisions of this sort had not been covered by the fluctuations clause although *ex gratia* allowances in this instance were made by the Government and recommended to public authorities.

In October 1967 a revised clause 31 was approved by the JCT. Firm price contracts had had the fluctuations clause deleted up to that point but the new clause was much more extensive, and contained optional sub-clauses A and B. Sub-clause A gave effect to full fluctuations and subclause B, applying to firm price contracts, permitted the reimbursement of costs arising from 'political risks', such as a contribution levy or tax payable by an employer of labour. Additional subclauses C and D which applied to both types of contracts were introduced and in July 1973 a subclause E came into effect.

The situation was further complicated in April 1973 when a productivity deduction from labour cost fluctuations under building contracts was introduced under Phase II of the Government's anti-inflationary measures. (This was removed from work executed on and after 1 August 1976 – see 4.3.1. below.)

In December 1973, after strong and continuous pressure from the industry, the Government advised that on a trial basis firm price tenders should normally only be invited where the contract period did not exceed one year and that for longer contracts fluctuations would apply. (The present position is set out in Chapter 9 of the Code of Procedures for Local Authority Housebuilding – see section 2.4.5.) This was coupled with the announcement that variable price (fluctuations) contracts, in the public sector would be adjusted on a formula based on indices of cost as recommended by the Economic Development Committee for Building. Local authorities were informed in a DOE Circular that this 'should help to achieve more realistic tender prices and reduce the administrative burden'.

An outline of this new basis, and a review of clause 31F (now 40) introduced into the Standard Form of Building Contract to give legal effect to it, will be found later in this chapter (4.4).Until this became fully operational local authority tenders on a 'fluctuations' basis were invited on clause 31A but in due course the formula method for public authority work became almost universal. In private sector contracts, either method of adjustment, 31A (now 39) or formula can be used, as the employer decides.

For 'firm prices', clause 31B (now 38) applies.

In the 1980 edition of the Standard Form the fluctuations clause number has been changed from 31 to 37-40 as detailed below. Future references in this chapter will therefore be to the new clause numbers.

The fluctuations provisions similar to those in the Standard Form are incorporated in nominated and non-nominated standard forms of subcontract and are dealt with in Chapters 5 and 6.

It is proposed to deal firstly with 'traditional' fluctuations (clauses 38 and 39) and, after noting the former productivity deduction and price code requirements, go on to explain the operation of the price adjustment formula and the provisions of clause 40.

For Government work, Form GC/Works/1 incorporates a clause giving effect to the price adjustment formula (see section 9.7.4).

4.2 Fluctuations – Clauses 37 to 40, Standard Form

An important change in the 1980 edition of the Standard Form is the transfer of former clause 31 (fluctuations) to the end of the contract (clauses 37 to 40).

The opportunity has also been taken to alter the format, in that the appropriate provisions only, published separately, will be incorporated in the contract. Formerly deletions had to be made since the terms of clauses 31A, B, C, D, E and F were all set out in full.

Clause 37 briefly states that fluctuations must be dealt with in accordance with one of three choices, to be set out explicitly in the Appendix (and by inclusion in the preliminaries in the bills, clearly indicated at the time of tendering). The contract conditions will then incorporate the clause relevant to the fluctuations provision chosen, ie:

(i) 'full' fluctuations – clause 39
(ii) fluctuations limited to contributions, levies, etc, payable by an employer as such (known as 'firm price') – clause 38
(iii) contract price adjustment by the formula method under formula rules – clause 40.

It is made clear that if clauses 39 or 40 have not been chosen, clause 38 must apply. (In the past it happened on occasion that all clauses were deleted.)

The range of fluctuations recovery under clause 39 has been extended, particularly in regard to labour costs, as the following pages show. It should be remembered that these provisions cover only the main contract work including sublet ('domestic subcontract') work under clause 19, in which the main contract fluctuations provisions in clauses 38 or 39, as applicable, must be incorporated (see non-nominated 'blue' form clause 25). The importance of non-nominated subcontract fluctuations terms being indentical to main contract terms is obvious apart from any contractual requirement.

Nominated subcontract work is covered by the terms of the appropriate nominated subcontract:

1980 edition of JCT Nominated Subcontract Form – NSC/4 or 4a, clauses 34–37
Pre–1980 edition of Nominated Subcontract ('green') Form – clause 23

It can, of course, happen that the fluctuations provisions for nominated subcontractors (which will be reflected in the adjustment of the PC sum) differ from those in the main contract (ie, the contract is let on a different basis). This, however, is generally undesirable since a main contractor on a firm price (clause 38) may consider it inequitable that one or more of his subcontractors should have the benefit of full fluctuations (see also section 2.4.5).

In the case of work carried out by the main contractor under clause 35.2 (work otherwise reserved for a nominated subcontractor) the fluctuations recovery would depend on the terms of the contract for that work.

In reviewing the contract provisions clause 39 (traditional fluctuations) will be considered first since this has a bearing on clause 38 ('firm price'). Several features common to clauses 38 and 39 will then be reviewed together to avoid duplication. Clause 40 is dealt with separately in section 4.5 below, after the principles of the price adjustment formula have been explained.

4.2.1. CLAUSE 39

There are three subclauses dealing with the main headings for recovery of additional costs.

39.1 relating to labour costs
39.2 covering contributions, levies, etc payable by an employer of labour
39.3 reimbursing increased costs of scheduled materials.

They are all prefaced by the provision that the contract sum shall be deemed to have been calculated in the manner set out, the basis of pricing then being identified (see also 38.1 and .2).

Labour Costs

At the date of tender, that is ten days before the date for receipt of tenders (clause 39.7.1), current wage rates and those notified increases already the subject of a published decision of the National Joint Council for the Building Industry, or of the appropriate wage fixing body, must be taken into account in pricing, together with existing rates of contribution, taxes or levies on labour which would be payable on these rates by an employer (eg, national insurance contributions at the present level on known wage increases).

If higher labour costs as defined result from a decision announced after the date of tender, the net additional costs are reimbursed to the contractor. 'Workpeople' for fluctuations recovery include not only those engaged on the works either on or adjacent to the site, but also those directly employed by the contractor who are neither on nor adjacent to the site but are engaged in off-site production of materials or goods for use in or in connection with the works. (The JCT recommends that procedures be agreed at the outset for verifying numbers, time spent, etc.) There are others employed on the site on supervisory or other work who are not defined as workpeople. These are now also covered under clause 39.1.3. Increases will be limited, however, to the amounts which would have been payable in respect of craftsmen. Normally this would be the NJC rate but if other rates apply (eg, the plumbing JIB) the relevant rate must be settled at the tender stage, and 'if more than one rate is promulgated' the highest must be used. There is a weekly qualifying period of not less than two whole working days on site (parts of a day do not count) with reimbursement pro-rata (clause 39.1.4). A similar principle applies to such payments as are due under clause 38 and to sublet work as specified under 39.4. The recovery provision under clause 39 extends to increased employer's liability and third party insurance premiums on wage rolls increased by pay awards (but not increased rates of premium). There is now provision for increases in travel allowances payable to workpeople under a joint agreement, including increased cost of employer's transport where used, (defined at the tender stage) or public transport fares. Emoluments and expenses including holiday credits payable by the contractor under NJC rules also qualify. This also applies to other employees, such as electricians, covered by some other wage fixing body.

Additional payments relating to a bonus calculated by reference to wage increases, which become due under an incentive scheme or a productivity agreement under NJC or other rules on which prices were based, are now covered under 39.1.1.4. The guaranteed minimum bonus is now widely accepted as mandatory and therefore recoverable within the scope of clause 39.1. In a joint RICS/NFBTE statement issued in October 1977 the view was expressed that on all future contracts guaranteed minimum bonus increases or decreases are recoverable by or from the contractor.

Employers' contributions, levies and taxes

Initially, this subclause (39.2) was a direct result of the introduction of Selective Employment Tax. The prices in the tender should include the (statutory) types and rates of contributions, levy and tax payable, or premiums receivable, by a person in his capacity as an employer of labour at the date of tender.

Even if at the date of tendering a future increase such as higher national insurance rates of contribution are known, it should not be included in pricing but claimed later as a fluctuation once it becomes operative. A claim would also arise where subsequent qualifying wage increases resulted in higher wage-related national insurance contributions or other levies, etc based on labour costs.

The introduction in April 1978 of the new system of state pensions under the Social Security Pensions Act 1975 necessitated amendments which appear in clause 39.2.7.

For the purpose of this subclause contracting out contributions must be classified as follows:

(a) *where contracting out is decided by the employer* – this will be ignored and recovery under the clause will be limited to the level of state scheme contributions

(b) *where contracted out employment is a result of decisions of a wage fixing body* (ie, under an Occupational Pensions Board certified private occupational pension scheme) – contributions are allowed in full (eg, plumbing JIB pension scheme) and will be claimable under clause 39.1.

This subclause (which does not cover Construction Industry Training Board levies) is drawn in very general terms and relates to all statutory payments which increase the cost to an employer of employing labour as defined in subclause 39.1.1 and 39.1.3. It provides cover for any new labour levy which the Government might in future introduce. Selective Employment Tax, which was not originally covered in the fluctuations provisions, subsequently came within the scope of this subclause, its reduction and ultimate removal resulting in claims on contractors for a reduction in the contract sum where tenders were based on the higher rate.

Materials
Reimbursement of extra costs covers materials, goods and electricity consumed on the site for the execution of the works including temporary installations. Fuels may be included where specifically stated in the contract bills. A list of materials, goods and electricity (and fuels where appropriate), setting out the market prices current at the date of tender on which prices are based, must be submitted by the contractor and attached to the contract bills. (It is undesirable to make this list too lengthy because of administrative costs.) The contract sum is then adjusted for any difference between market prices at the date of purchase and those set out in the tender list.

Market prices include duties or taxes imposed by parliament such as import and sales taxes (VAT, which the contractor can treat as input tax, is not included). This provision includes timber used in formwork but does not cover consumable stores, plant and machinery (clause 39.7.2).

4.2.2 CLAUSE 38
This clause applies to firm price contracts and is limited in scope. Subclause 38.1 relating to contributions, levies and taxes payable as an employer, although somewhat similar to subclause 2 of clause 39 in that it extends to workpeople on and off site, and to site supervisory staff as defined, is confined to actual statutory rate changes (ie, in national insurance, only the effect of an increase in rates of contributions).

The new state pension scheme, operative from April 1978, necessitated a special provision in clause 38.1.8. Where contracting out takes place the extent of the recovery on increased contributions is in every instance limited to the level of contributions under the state (non-contracted out) scheme.

Under clause 38.2 (materials, etc), the fluctuations payments are limited to those materials, goods and electricity (and to fuels where specifically stated in the bills – see 39 above) included in the list submitted by the contractor with the contract bills at the time of tendering, and only to the extent that costs have increased or decreased owing to a change in duties or taxes by virtue of any act of parliament becoming effective at any time after the date of tender. (VAT is not included where the contractor can treat it as input tax.) Changes in costs due to currency adjustments, for example, are not covered.

4.2.3 CLAUSES 38.3 AND 4 AND 39.4 AND 5

Seven important points must be noted in the operation of these sub-clauses: which are considered together to avoid repetition.

(a) The 'date of tender' is ten days before the date for the receipt of tenders (clauses 38.6.1 and 39.7.1).

(b) Clauses 38.4.5 and 39.5.5 require the evidence and computations of amounts actually disbursed which are necessary to check all amounts under 38 and 39 to be provided by the contractor to the architect or quantity surveyor. Particular reference should be made to the recent inclusion of electricity (and fuels where specifically provided in the bills) in the materials fluctuation provisions under 38 and 39. In regard to those employed on site other than 'workpeople', a weekly certificate signed by the contractor will be required as evidence of the validity of the claim (including 'domestic subcontractors' employees).

(c) Amounts payable or allowable under clauses 38 and 39 must be added to or deducted from the contract sum (clauses 38.4.4 and 39.5.4). Under clause 3 this should be included in the next interim certificate without any addition for profit (clauses 38.4.6 and 39.5.6) but without deduction of retention (clause 30.2.2.5).

(d) The fluctuations conditions in the main contract must be incorporated in contracts for work sublet under clause 19 (clauses 38.3.1 and 39.4.1) (see clause 25 non-nominated 'blue' subcontract form – Chapter 5). The main contract percentages for 38.7 or 39.8 should apply to these subcontracts, but if a higher percentage is inserted the main contractor would have to bear the excess. In the case of nominated subcontractors or suppliers fluctuations will be determined by the appropriate subcontract or conditions of sale (clause 38.5 and 39.6). Where under clause 35.2 a contractor carries out subcontract work, fluctuations would depend on the terms of his accepted tender.

(e) Written notice of increases in labour and materials costs must be given promptly to the architect (clause 38.4.1 and 39.5.1) as a condition precedent to payment (clauses 38.4.2 and 39.5.2).

(f) Fluctuations do not apply to daywork under clause 13.5.4 since labour would be charged at the current rate, nor does it apply to changes in VAT (clauses 38.5 and 39.6).

(g) It is within the discretion of the quantity surveyor and the contractor to agree fluctuations payments (eg, a lump sum to avoid detailed calculations) (clauses 38.4.3 and 39.5.3).
Note: This would appear to be outside the scope of the arbitration provisions of Article 5 – see section 3.10.).

4.2.4. CLAUSES 38.7 AND 39.8

All fluctuations were net until the introduction of this subclause in July 1973. These clauses stipulate an optional percentage addition as set out in the Appendix to the contract, to provide for additional recovery on defined fluctuations. The tendering documents should give the figure to be inserted. All too often it is 'nil' although the JCT discussed 20 per cent as a figure but did not include it in the clause itself: an unfortunate omission in the light of subsequent experience. (In a 'Guidance Note' dated 31 July 1974 the NJCC considered that 20 per cent under clause 31A and 10 per cent under clause 31B would be realistic).

Present Government policy (at late 1979) is that an addition of up to 15 per cent may be allowed in local authority contracts on payments under clause 39 but no addition under clause 38 (see section 2.4.5.)

It should be noted that the inclusion in the 1980 edition of supervisory, etc staff in the definition of labour for recovery of increases under clause 39 may have some influence on the percentage under 39.8 which might normally be considered to reflect inter alia increased supervision costs not otherwise recoverable.

4.2.5. LIMITATION ON RECOVERY OF FLUCTUATIONS

After the completion date set out in the Appendix, as extended under clause 25 (see Chapter 3) or clause 33 on war damage, certain increases (or decreases) in costs may arise under the fluctuations provisions (clauses 38 or 39). A new amendment (clauses 38.4.7 and 39.5.7) provides that these additional costs may not be added to the contract sum where clause 25 operates without amendment and the procedures have been fully observed – ie, the architect has fixed in writing the later completion date he considers appropriate in respect of every written notification by the contractor, who must give the architect all the information he reasonably requires. This means that fluctuations payments are 'frozen' at the level established at the date finally fixed for completion. This brings the operation of 38 and 39 into line with the principle of 'freezing the indices' under clause 40.7 (see 4.5 below) which has operated since March 1975.

Nominated subcontracts NSC/4 and 4a contain similar provisions in clauses 35 and 36. Thus where the nominated subcontractor is not late on his contract, the fluctuations payments due to him are not affected even if the main contractor is late. Similarly, if the nominated subcontractor fails to complete on time and the architect has certified accordingly, his loss of right to recover fluctuations after that date will not affect the main contractor's entitlement.

4.3 Contract Payments and Government Policy

From 1973 to 1978 payments under the standard conditions of building contracts were affected by the exercise of statutory and other powers by Governments in support of their prices and incomes policies. The following pages review this significant development. It could be dismissed as history – and one can only hope that history will not repeat itself. Government intervention in commercial contracts freely entered into, can undermine sound business relationships.

4.3.1 PRODUCTIVITY DEDUCTION – HISTORICAL NOTE

The productivity deduction ceased to apply to the reimbursement of defined additional labour costs on building work executed on and after 1 August 1976. No one lamented the disappearance of this illogical and confusing provision which substantially reduced the contractual recovery of labour fluctuations.

Although this is now in the past it was an important factor in contract work from 1973 to 1976. It may be useful therefore, as a matter of record, to describe briefly these provisions under the Counter Inflation Act 1973, designed 'to ensure that the benefits of increased productivity are passed on to the consumer'. It was an early phase of continuing Government policy to seek to exercise some control over prices, the more recent provisions being reviewed in the next section.

Under the fluctuations clauses in the standard forms of contract (main and subcontracts) there was a right to recover defined additional labour costs. The statutory provisions on productivity deduction reduced this right. Increased material costs were not affected provided they were justified under the prices code.

Labour costs were statutorily defined but certain items which were included in this definition were not subject to productivity deduction: eg, increases in national insurance conributions, training costs, equal pay commitments and occupational pensions provisions.

Each firm had to calculate and allow its own productivity deduction percentage by reference to the firm as a whole, the total labour cost being expressed as a percentage of the firm's total costs. The initial provision in Stage II (see below) was a 50 per cent deduction from labour cost increases where labour costs did not exceed 35 per cent of total costs. Above this the productivity deduction was scaled down. This was continued in Stage III but in Stage IV the productivity deduction was reduced to 20 per cent where labour costs were between 15 per cent and 35 per cent of total costs. The scale in Stage IV covered a wider range. This continued until 1 August 1976 when the productivity deduction no longer applied.

There was provision for relief from the deduction where, for example, a contract was running at a loss or the firm as a whole was currently in a loss situation.

The statutory references to productivity deduction in orders under the Counter Inflation Act 1973 are as follows:

		Productivity deduction from
Stage II	Price and Pay Code Order 1973 – SI 657: 1973, para 29	29 April 1973
Stage III	Price and Pay Code No. 2 Order 1973 – SI 1785: 1973, para 32	1 November 1973
Stage IV	Price Code Order 1974 – SI 2113: 1974, para 34	20 December 1974 (transition from 1 November 1974)
Stage V	Price Code Order 1976 – SI 1170: 1976, para 71	ceased 1 August 1976

The cessation of the productivity deduction made it no longer necessary to include in the indices for the price adjustment formula a labour component on which the deduction had been calculated.

However a sanction was introduced in paragraph 43 of the Counter Inflation (Price Code) Order 1976 (SI 1170:1976) where remuneration was in excess of limits prescribed in Section 1 of the Remuneration Charges and Grants Act 1975. This resulted in emergency amendments to the standard forms of contract in 1975 but as explained in the next section the situation was materially changed when the 1975 Act lapsed on 31 July 1978.

4.3.2 PRICE CODE – ADDITIONAL CONDITIONS

After 1975 and until 1978 there were two phases of Government pay and prices policy in relation to building contracts. Initially there was a legal sanction enforceable under the Remuneration Charges and Grants Act 1975. This act was allowed to lapse on 31 July 1978, and thereafter reliance had to be placed on contract conditions to ensure the obsevance of pay policy which from 1 August 1978 had no legal sanction whatsoever.

Whereas the 1975 restraints were enforced by local authorities as well as Government in contract conditions, after 1 August 1978 when there was no longer statutory backing local authorities refrained from introducing contract conditions similar to those in Government contracts to ensure the observance of incomes policy by contractual terms alone.

Several aspects of these conditions are briefly considered below.

On 1 December 1975 an emergency amendment (1/1975 Additional Clause 30A) was published by the JCT as an addition to, but not incorporated in, the Standard Form of Contract, with minor amendments to apply to theStandard Form with Approximate Quantities and the Fixed Fee Form. There were similar amendments to the subcontract forms: 'green' (clause 11A) and 'blue' (clause 13A). (*Note*: A comparable provision was added to GC/Works/1 – Supplementary Conditions 163/164 – see Chapter 9.)

This was requested by the Government in pursuance of amendments to the Price Code under the Counter Inflation Act 1973, as amended by the Remuneration Charges and Grants Act 1975, which were statutory provisions overriding contract conditions.

In brief, an employer could refuse to pay labour fluctuations in contracts, or increases in labour prime cost in daywork, where after 11 July 1975 that payment was in respect of excess remuneration under the code or was a part of excess remuneration (ie, the entire payment was not recoverable).

To check on the position public authorities were asked by the Government to include in their contracts a requirement on contractors and subcontractors to certify that daywork and fluctuations payments excluded excess remuneration.

Regular prompt certificates in a prescribed form were required from contractors and similar requirements were placed on subcontractors (nominated and non-nominated), the contractor to pass their certificates on to the architect. Although there was no requirement to use this amendment in private work, the provisions of the code applied equally. Where the price adjustment formula was used (clause 31F) a certificate was still needed for daywork. In preparing index figures, however, only NJC increases which were, of course, within the code, were used and observance was automatic.

The policy situation became less clear from 1 August 1977. From that date a breach of the TUC statement of 22 June 1977 supporting the '12 month rule' was regarded as a payment in excess of the current pay policy and certificates under clause 30A were then in respect of the period between pay rises. Payments in excess of the Government 'guide lines' of 10 per cent were outside the scope of the clause from that date, the Government having declared its intention to deal with these breaches by sanctions, such as withholding Government contracts.

The removal of legal support, with the lapse of the Remuneration Charges and Grants Act 1975, resulted in these clauses being excluded from contracts let after 1 August 1978 under the standard forms. From that date certificates under clauses in existing contracts were no longer required. (On the same date GC/Works/1 Supplementary Conditions 163/164 were also withdrawn.)

On Government contracts, however, tougher measures were introduced in the spring of 1978 (GC/Works/1 Supplementary Conditions 185, 186 and 187), and despite the loss of legal sanctions these clauses remained in operation until 14 December 1978, when after a Government defeat they were withdrawn. Since these conditions were almost entirely confined to Government contracts they are briefly referred to in Chapter 9. The only application to the Standard Form was the unilateral introduction of a supplementary clause 30X by the Department of Health and Social Security, primarily for hospital contracts, which was also withdrawn in December 1978.

4.4 Price adjustment formula

The concept of the variable price contract is that if there are specified increases in the cost of labour, materials, etc after the tender price has been prepared, the defined extra costs falling on the contractor are to be added to the contract sum.

The long standing price fluctuations clause has certain drawbacks – the question of definitions, arguments on what is, and what is not allowable, shortfall in recovery (eg, in preliminaries, plant, overheads), delays in payment, and extensive work in documenting and checking claims. Alternative provisions to give effect to price variations on a formula basis necessitate a valuation of work executed each month by reference to national index figures.

The old idea of using index figures has now been presented in a more sophisticated and more complicated form. In essence the idea is simple. The index figures of relevant costs at the time of tendering form the base. Subsequent monthly increases or decreases in the index figures result in percentage adjustments to the valuation of work as it is carried out, month by month. Individual prices and actual extra costs on specific contracts do not enter into the calculation. In its simple form it is a very blunt instrument; in its new, sophisticated form it is hoped that its finer adjustments will justify the work and expense involved.

The new formula adjustment, which came into effect under the Standard Form of Contract, where agreed, from 1 May 1975, was the result of extensive work by the Economic Development Committee for Building, which published a report in 1969 and detailed recommendations in October 1973. The principle has applied to civil engineering contracts since mid 1973 and was introduced into GC/Works/1 early in 1974. Clause 31F (now 40) was incorporated in the Standard Form (with quantities) to give effect to these provisions as an alternative to clause 31A (now 39) which continued to operate where so decided. A change in the method of reimbursement during the currency of a contract is not feasible.

4.4.1 BASIC PRINCIPLES OF THE FORMULA

The methods to be used in operating the formula are set out in formula rules published by the Joint Contracts Tribunal as listed below.

These rules are given contractual effect, by incorporation by reference, in the appropriate clause of the main or subcontract (eg, Standard Form 40). There are two series as follows:

Series 1
34 work category indices:
Formula rules (main contract) – dated 3 March 1975
Formula rules (subcontracts) – dated 3 May 1975

Series 2
48 work category indices:
Formula rules (main contract) – dated 4 April 1977 (with 1980 amendments)
Formula rules (subcontracts) – dated 4 April 1977 (revised 1980 see below)

Series 2 formula rules (main contract) are in three sections:
(1) definitions, exclusions, correction of errors, etc
(2) operation of work category and work group methods
(3) application of specialist formula to main contractor's specialist work

The 1980 amendments are largely consequential, reflecting in particular changes in clause references eg, clause 13 (variations), clause 30 (payments) and clause 40 (use of price adjustment formulae).

In the subcontract formula rules specialist engineering indices are available for installations for electrical work; heating, ventilating and air conditioning; lifts; structural steel; and also in Series 2, for catering equipment installations. These have indices for labour and materials which cover the prices for specialist work, nominated or non-nominated. The Nominated Subcontract Formula Rules Series 2 were substantially revised in 1980 to accord with the JCT Nominated Subcontract Conditions (NSC/4 etc). In addition Part II has been transferred to Part III and aligned with Part III of the Main Contract Formula Rules (specialist engineering work). Series 2

Formula Rules (4 April 1977) will still be published for the 'green' and 'blue' subcontract forms.

At the same time, with the appearance of the JCT Design and Build Form (see section 7.5), an opportunity has been taken of adapting the formula rules for use with this form. All three are now published by the JCT in one document as follows:

(1) Standard Form of Building Contract Formula Rules – Work Category Indices Series 2 – dated 4 April 1977 (revised 1980).
(2) JCT Standard Forms of Subcontract – NSC/4 and NSC/4a – clause 37 – Nominated Subcontract Formula Rules (1980).
(3) Formula Rules for use with the Standard Form with Contractor's Design – 1980.

The principles set out in the formula rules are considered below, work categories and indices being dealt with in the next section.

The first step in operating the formula is, before inviting tenders, to allocate as far as possible all the items in the bills of quantities to the appropriate work categories. The exceptions are preliminaries, prime cost and provisional sums and work covered by specialist engineering indices.

The value of work carried out in each category in respect of each (monthly) valuation period will be increased by reference to the percentage increase for each category of the appropriate index figure current at the mid point of that valuation period, over the firm index figure for the base month on which the tender was based.

The significance and main features of the 48 category series introduced in July 1977 are considered in the next section 'Work Categories'. The basic principles are unaffected but the extent and method of compilation of the indices differ. The comparative table in Appendix G at the end of this book will assist in an appreciation of the changes made. Contracts already let on the 34 category series must continue on this basis since there is no interchangeability.

Where tenders are invited, the base month is given, normally the month prior to that on which the tender is returned. Therefore the prices in the tender should relate to the indices for that calendar month. This time interval, and the base dates used, should be borne in mind when tendering. The Series 2 indices are published in the first half of the following month. Labour rates based on NJC decisions, etc are those payable on the 15th of the month, material prices in the preceding month and plant costs in the month to which the category index refers.

Interim valuation for each work category takes place each month and the value of work in that period is subject to adjustment. Since the index current in the mid-point of the valuation period is related to that of the base month, the date of valuation thus assumes a new significance.

The amount recoverable in Government contracts is 90 (previously 85) per cent of the increase calculated. Ten per cent is regarded as a non-adjustable element to be stated in the tender documents. This provision is also included in contracts where central funds are involved, that is, in almost all public contracts, in conformity with Government policy. It represents an important firm price element to be taken into account when tendering. In private work there is no provision for the non-adjustable element in clause 40 (see below).

Valuations apply to measured work and will be in the work categories listed (if limited to work groups, the relevant indices will be weighted according to tender prices). Interim valuations (and the allocation to work categories) should be as accurate as possible, particularly when prices are rising steadily. Apart from correcting incorrect allocations to work categories and amending calculating errors (quantity surveyor's calculations to be available to the contractor on request), there should be no subsequent adjustment of valuations. Adjustment will be needed if the final index figure for the base or any month (published three months thereafter) is different from the earlier provisional one. Excluded from the formula adjustment are:

(a) *PC sums:* Since the use of the formula is now agreed before nomination (see NSC/1 Schedule 1 Appendix F), nominated subcontractors would be covered by the appropriate work category indices or by one of the special indices for specialist engineering installations. Terms of sublet work will determine the use of the formula. The appropriate subcontract clauses are reviewed in Chapters 5 and 6.
(b) *Nominated suppliers*: price ruling per conditions of sale. ('Fix only' work may be adjusted on special indices as provided in the bills.)
(c) *Unfixed goods and materials:* included at delivery price in interim certificate but subject to formula once incorporated in works. (Rapidly rising market prices raise interesting possibilities.)
(d) *Daywork:* based on agreed rates (clause 13.5.4).
(e) *Variations:* if based on bill rates these would be allocated to work categories, but other work on specially agreed rates would be excluded.
(f) *Retention money*: full value is already adjusted in monthly valuation so that price increases are subject to retention.
(g) *Articles manufactured outside UK:* see clause 40.3 below.
(h) *Amounts due under clauses 26.1 and 34.3*
(i) *Credit for old materials.*
(j) *Work specifically excluded:* based on actual cost; or executed on a specially agreed basis.

Three points should be specially noted:

(i) The 'balance of adjustable work' such as prelims, water, attendance and insurances is not allocated to work categories but is adjusted by the same percentage as the total of adjustable work included in the indices. This adjustment factor should be calculated on acceptance of the tender.
(ii) The balance of measurable work in interim certificates issued after the month in which the certificate of practical completion is issued, is adjusted pro-rata to the prior overall total interim payments figures.
(iii) Work after the contract completion date (extended where approved) is

adjusted for the appropriate categories on the index figures effective at the mid point in the month when the work ought to have been completed – ie, the indices are 'frozen' (see clause 40.7).

4.4.2 WORK CATEGORIES

One of the basic requirements of formula adjustment is the division of work in the bills of quantities into work categories.

In Government contracts, and in others where stated in the bills, 'work groups' (13) are used, not 'work categories' (34 for Series 1;48 for Series 2 – see Appendix G). Bills of quantities must reflect these divisions and a schedule to the bills, or an annotation of the items, must show the work category within each group in which bill items fall. Once the tender is accepted, the contractor will be bound by the allocation to categories of the bill items. The 'balance of adjustable work' should be as small as possible.

If there are not bills of quantities, prior to the award of the contract, the lump sum tendered must, if possible, be split between the various work categories if adjustments on this basis are to apply. The JCT pointed out that the 'Without Quantities' Form would have to be adapted in the unlikely event of the formula being used in these circumstances.(section 7.5)

Paragraph 18 of JCT Practice Note, 20 refers to the application of the formula where bills of approximate quantities are used.

The monthly *Bulletin of Construction Indices* published by the Property Services Agency through HMSO for Series 1 gave due weight on the basis of the analysis of 50 representative bills of quantities to the labour, plant and materials components in the prices of 34 work categories. A full explanation of the formula method and the compilation of indices was given in the NEDO booklets *Guide to Practical Application of the Formula* and *Description of the Indices*. These publications were by way of explanation and did not form part of contract conditions. After pressure for independent monitoring of indices, the Government set up a standing committee representing the professions, the industry and the employers.

Even in 1975, when the system was first introduced for use with the Standard Form, the view was expressed that formula adjustment for specialist work not covered by the specialist engineering formulae was not fine enough, particularly if related to only one or two categories out of the 34 provided. It was felt in particular that the appropriate indices did not fully reflect the balance of labour, materials and plant used.

In June 1975, therefore, the National Consultative Council's Standing Committee on Indices for Building Contracts started a review of the 34 categories. The result was the publication in early 1977 of a new Series 2 of 48 categories to apply to contracts let from July 1977 onwards, the JCT Formula Rules Work Category Indices, Series 2 being dated 4 April 1977 (JCT Practice Note 23). These two series (34 and 48) are not interchangeable and contracts let on the 34 category indices will continue on this basis. The new indices published from April 1977 onwards, with June 1976 as the base of 100, were built up from extensive data provided by many trade associations. A Series 2 booklet has been prepared to replace the original *Description of the*

Indices booklet and it gives full information on the compilation of the new indices and work categories. To assist in allocating bill items to the appropriate categories, the new booklet gives details of the SMM sections for each category. A Series 2 *Guide to application and procedure* has also been published. (see Appendix I for publications available.)

The table in Appendix G compares the two series as far as possible and broadly indicates the appropriate SMM sections in the 5th and 6th editions for each of the 48 categories. Experience has apparently shown that on average contracts 27 of the 34 categories have been used. It is estimated that 32 of the 48 categories will be used on average in the new series. For scaffolding subcontractors two separate indices are now provided – for normal scaffolding contract work and for hiring.

4.5 Standard form clause 40: price adjustment formula

To give legal effect to the use of the price adjustment formula a new clause 31F (now 40) was published by the JCT together with formula rules and Practice Note 18 on 24 March 1975. One or two minor amendments were made to the clause in April 1977 when Series 2 was introduced and Practice Note 23 followed in November 1977 (see Appendix C).

As already indicated, clause 37 is a brief one, enumerating three choices for dealing with fluctuations, the one selected to be set out in the Appendix to the contract and communicated to tenderers through the preliminaries in the bills. The appropriate clauses would then be incorporated in the contract conditions.

Where clause 40 applies, bills must be prepared so as to supply all essential information including annotation to show work:

(i) allocated to work categories
(ii) of main contractor covered by specialist formulae
(iii) not subject to formula adjustment
(iv) balance of adjustable work
(v) 'fix only' or covered by a provisional sum.

Note: The contract provisions in GC/Works/1 are dealt with in Chapter 9 and subcontract provisions in Chapters 5 ('green' and 'blue' forms) and 6 (JCT Nominated Subcontract Form).

Clause 40 must be read in conjunction with the formula rules current at the date of tender (current issue, Series 2, dated 4 April 1977 – as amended 1980). They form an integral part of the clause and as part of the contract conditions they are contractually binding. Adjustments under this clause must be to sums exclusive of VAT (subclause 40.1.1.2).

Subclauses 1.3 and 1.4 provide that the formula adjustment must be effected in all certificates of payment, and that rule 5 adjustments (correction of errors on previous certificates) must be included in the next certificate. Since there will be a regular monthly valuation, adjustment and payment, the provision in clause 30.1.2 for valuations 'where necessary' is deleted by subclause 2. The quantity surveyor and contractor may, by altering the method and procedure, agree amounts payable under the clause (other than amounts due to subcontractors) provided the amount due is substantially the same as under the rules (subclause 5). The retention percentage is applied to adjustments under clause 40 (see clause 30.2.1.1). It does not apply to adjustments under clauses 38 and 39 (see clause 30.2.2.5).

Where articles in a list attached to the contract bills which are manufactured outside the United Kingdom show any change from the market price in sterling at the date of tender (including import duty, sales tax etc but excluding VAT) there is an appropriate adjustment (subclause 3).

In regard to nominated subcontract work, the main contract adjustments will not apply. The relevant indices apply independently as set out in the subcontract NSC/4 and 4a clause 37 (ie, specialist formula where applicable, otherwise appropriate work categories if relevant; failing that, some other agreed method). In sublet work, unless the contractor and non-nominated ('domestic') subcontractor agree to the contrary, the appropriate specialist formula or work categories will be used (subclause 4.2). (See Chapters 5 and 6 for subcontract provisions.)

Should the Monthly Bulletin containing the indices be discontinued or unavailable, adjustment under the clause is to be 'on a fair and reasonable' basis. If publication is resumed before the issue of the final certificate, adjustments will be retrospective (subclause 6).

The Appendix to the contract sets out the base month and whether the work category or work group method is being used. In the local authority edition only, the non-adjustable element, not to exceed 10 per cent, must be inserted. (*Note*: This was reduced from 15 per cent in 1976 – DOE Circular 19/76 of 10 February 1976; July 1976 Standard Form Amendment Q12/1976.) In the private edition there is no non-adjustable element.

'Freezing' the operation of the formula was first introduced in GC/Works/1. Effect was given to this principle in the Standard Form when clause 31F (now 40) was incorporated, subclause 7 providing that if the contractor fails to complete the works by the completion date (the original date for completion as extended under clause 25), the indices applicable to the valuation period in which work ought to have been completed (ie, the completion date as defined under clause 25) will be applied to adjustments in all subsequent interim certificates.

This is subject to clause 25 being unamended and the architect having already given in writing his decisions on all applications for an extension of time. An immediate reference to arbitration on this latter point may be made.

It should be noted that in the 1980 edition of the Standard Form the principle of this provision has now been extended to clauses 38 and 39. In applying these restrictions to sums due to nominated subcontractors, currently under NSC/4 or 4a clause 37.7 the relevant date for freezing is related to an architect's certificate under clause 35.15 of the main contract.

4.6 Assessing the new system

It is by no means easy to assess the significance of these provisions. Tendering initially was a little more hazardous and expensive than under clause 31A (now 39) until contractors and quantity surveyors became used to the system. Smaller firms are still finding it hard to cope effectively with the complicated considerations.

In addition to the non-adjustable (ie, firm price) element already referred to, the national indices may not be fully comprehensive and may not reflect individual contractor's costs, such as labour costs including bonuses, and the local purchase price of materials compared with the national index price. This aspect needs constant review and the new 48 categories endeavour to meet some of the criticisms although adding to the complications of the system. Furthermore the proportions for labour, materials and plant (ie, the work mix) in compiling the indices may differ from that of the actual contract and this makes the formula unacceptable where work is unduly labour intensive. Cash flow must also be watched. Pricing in tenders, if incorrect or too low, tends to be perpetuated.

On the other hand, in comparison with clause 39, there is more effective cover for increases in all materials, plant and labour costs. Preliminaries, overheads and profits also benefit from adjustment. The NFBTE Sensitivity Studies are valuable in assessing the operation of the formula (see Appendix I). Many felt that such a radical change should have been implemented gradually so that knowledge and experience could be used to ensure its success. The use of the formula on contracts of 12-24 months duration was regarded as a useful first step with contracts covering longer periods being included later.

After some five years of operation the general opinion appears to be that, once the formula method is fully understood, three benefits accrue from using clause 40 instead of clause 39:

(a) there is a considerable saving in clerical effort, although the formula adjustment calls for more highly skilled staff
(b) arguments on interpretation are much reduced
(c) the rate of recovery of fluctuations is higher under 40 than it is under 39 and cash flow is improved.

Experience has shown that to benefit fully from this a thorough understanding of the price adjustment formula is necessary. This chapter only gives an outline of the system and the publications listed in Appendix I should be carefully studied.

5

SUBCONTRACT CONDITIONS NFBTE/FASS/CASEC FORMS

The main contractor may decide or may be instructed by the architect to subcontract some of the work.

The fundamental point is that the main contractor is responsible for the whole contract and any subcontract work is to be carried out under his directions. The architect's instructions to the subcontractor are given through the main contractor, who is responsible to the architect and the employer for the satisfactory execution of the contract works. Under the conditions of contract there is no contractual relationship between the employer and subcontractors which, if desired, would have to be created by a separate agreement. If anything goes wrong the remedy is against the main contractor and it is for the main contractor to seek his own remedies against his subcontractors.

5.1 Distinctions between nominated and non-nominated subcontractors

Subcontractors on a labour and materials basis fall into two groups under the Standard Form of contract. Until January 1980 the two subcontract forms (nominated and non-nominated) were the concern of the NFBTE, FASS and CASEC as follows:

(i) The Standard Form of Subcontract (green) issued under the sanction of the NFBTE and FASS and approved by CASEC (revised April 1978) was used for subcontractors who were nominated by the architect under former clause 27. This incorporated the subcontract requirements and provisions of clause 27 of the pre-1980 Standard Form.

(ii) The Non-nominated Subcontract Form, the 'blue' form, approved by NFBTE, FASS and CASEC, published in 1971 (revised July 1978) and substantially amending the conditions originally published by the NFBTE is used for subcontractors chosen by the contractor under clause 17 of the pre-1980 Standard Form (clause 19: 1980 edition), approved but not nominated by the architect. This is known as subletting. (See later for information on a shortened version of this form.)

NOTE: (a) both forms have an appendix giving contract details of programme, dayworks, payment, etc.
(b) for the position under Scots law see section 3:1

From January 1980 the contractual arrangements are radically altered as far as nominated subcontracts are concerned.

The terms of reference to the Joint Contracts Tribunal were extended in June 1966 to enable it to assume responsibility for the production of a Standard Form of Nominated Subcontract, the specialist organisations of FASS and CASEC then becoming constituent bodies of the JCT for the first time.

After very extensive discussion, negotiation and drafting the JCT Nominated Subcontract Form was finally agreed and published in January 1980. The documents fully detailed in Chapter 6 are Tender NSC/1; Agreement NSC/2 or 2a; Nomination NSC/3; Subcontract NSC/4 or 4a. These supersede the Standard Form of Subcontract (the 'green' form) published by NFBTE, FASS and CASEC in respect of future contracts let on the 1980 form, but the non-nominated (blue) form will still operate for sublet work.

Since many contracts were current when the new forms were published with subcontractors working under the 'green' and 'blue' forms this Chapter deals with both forms examining, explaining and comparing their provisions. For the sake of simplicity it is written in the present tense but the situation referred to in the preceding paragraph should be borne in mind. Although the 'green' form will be phased out, it will be available for some time yet for use in connection with contracts under the pre-1980 Standard Form. It must not be used under the new 1980 Form.

The following Chapter 6 is entirely devoted to the JCT Nominated Subcontract Form, thus enabling the new terms to be separately studied and appraised. Comparisons are also made between the 'green' form and the JCT form to assist in an appreciation of the significant changes.

In addition, since this new form has necessitated substantial amendments to the main Standard Form of Building Contract, a separate section of Chapter 6 is devoted to the terms of clause 35 dealing from the main contract point of view, with major changes in nominated subcontract provisions and procedures. The references to the main contract in this present chapter are to the pre-1980 edition. In the new 1980 edition the major provisions relating to nominated subcontractors are, as already indicated, contained in a new clause 35.

The essential difference between nominated and non-nominated subcontractors is immediately apparent on an examination of the 'green' and 'blue' forms. In the former the architect plays a significant role, whereas in the latter his dealings are almost entirely with the main contractor particularly in relation to the contract dates, payments, etc. There are frequent references to the powers and duties of the architect in the nominated form and few in the non-nominated form. In the latter, apart from some incidental references, the architect is only mentioned in connection with the non-nominated subcontract works to the following extent:

(i) execution and completion to his reasonable satisfaction (clause 2)
(ii) instructions in relation to work (clause 8) and on making good defects

(clause 11 (2))
(iii) access to work (clause 17).

It should be remembered that whereas the use of the nominated form is supported by clause 27 of the Standard Form, the non-nominated form is used if both parties agree. The golden rule is that the agreed conditions for the industry should be used on every occasion without amendment.

5.2. Nominated Subcontractors

The architect decides, generally when tendering documents are being prepared, what work he wishes to have carried out by specialist firms. He then estimates the cost of this work and includes a PC sum in the bills of quantities with a suitable description, giving the name of the firm nominated, if known. PC sums, under which a nomination takes place, can also be created by the expenditure of a provisional sum (*cf* 1980 Standard Form clause 35 – see section 6.3.1.).

It can happen that for specialist engineering services etc, particularly where there is a separate design element in the work, the architect may make arrangements with a specialist subcontractor before the main contract is let. Care would have to be taken to see that any specialist programme eventually fitted in with the programme of the main contractor subsequently appointed, because he could reasonably object if it were at variance with his requirements.

On tendering, the contractor in pricing the bills should add a percentage to the PC sum for profit, etc. This is adjustable pro-rata to the PC adjustment in the final account under clause 30(5)(c) of the pre-1980 main contract. He also adds an amount for general attendance (eg, use of contractor's scaffolding, roads, etc, the provision of storage, accommodation, light, water, etc) which must be given as a separate item in the bills. There must also be fully described in the bills of quantitities for pricing, any other (special) requirements (eg, special scaffolding and access, unloading, hoisting, storage, power) detailed as separate items (SMM 5th edition B20; 6th edition B9). These priced items would only be adjustable on a change in requirements, etc under an architect's instruction.

In considering the question of nomination from the main contract standpoint the following pages refer to the pre-1980 Standard Form and to provisions and procedures prior to the introduction of the JCT Nominated Subcontract Form. Clause 27 of the pre-1980 Standard Form sets out the provisions where under a PC sum either included in the bills or resulting from the expenditure of a provisional sum a person is nominated by the architect to supply and fix materials or goods or to execute work.

The subcontract sum is to include a cash discount of $2\frac{1}{2}$ per cent for the contractor on all payments made by him to the nominated subcontractor within 14 days of the receipt of an architect's certificate.

The architect should, in good time, invite tenders from suitable firms for work which is the subject of a PC sum and make clear to firms invited that they will be nominated subcontractors. Once the architect has selected a suitable firm, he nominates, instructing the contractor to place an order for the work. If a specialist subcontractor's tender to the architect is for 'design and install', care must be taken to ensure that the design element in no way forms part of the subcontract with the main contractor.

If the architect is late with his nomination, the contractor might suffer loss or expense, and could claim under clause 24 of the Standard Form.

The architect must not nominate a subcontractor against whom the contractor has a reasonable objection and this would include failure to provide for the prescribed cash discount. If a contractor places his order on this basis without objection, he would lose his right to the 2½ per cent. This right of reasonable objection must be exercised before the contract is entered into. Grounds for objection might include an unfortunate previous experience with the firm, doubts about its capacity or financial stability (to be handled warily – and orally!), the introduction by the subcontractor of special conditions or an unwillingness to accept the main contractor's programme.

The architect may not nominate a subcontractor who refuses to enter into a subcontract covering the undernoted stipulations in clause 27a (i) to (x) which are all incorporated in the 'green' subcontract form:

- (i) satisfactory completion of work according to instructions and directions.
- (ii) compliance with the main contract insofar as it relates to the subcontract works
- (iii) indemnity for contractor against liability to employer
- (iv) claims arising through the negligence of the subcontractor
- (v) completion and provision for extensions of time
- (vi) damages to main contractor for failure to complete on time
- (vii) cash discount of 2½ per cent for payment within 14 days of receipt of architect's certificate
- (viii) retention and contractor's fiduciary interest as trustee for the subcontractor
- (ix) right of access for the architect
- (x) determination of subcontract where main contractor's employment is determined.

The following points should be noted:

- (a) Invitations to tender by the architect should clearly inform subcontractors of these terms.
- (b) The contractor should see to it that the terms on which he is asked to place his order do not differ in any respect from those set out in this clause, otherwise he can object to the nomination.
- (c) The nominated subcontract ('green') form incorporates those terms but clause 27 does not actually require its use.

If, despite reasonable objection, the architect insists on the order being placed with the subcontractor first nominated, this could be challenged under clause 2(2). If the response were unsatisfactory, and the matter not taken at once to an arbitrator, the contractor could place on record his view that the instruction, if pressed, would be in breach of the contract, any damage suffered to be the subject of a claim. If the contractor is then still required to place an order this might be regarded as an architect's instruction, so that any extra costs resulting could be claimed as an addition to the contract sum.

Subclauses (b) to (g) of clause 27 deal with the following matters from the standpoint of contractor and employer.

The architect is to inform the contractor of amounts included in certificates for payment to nominated subcontractors who should be notified direct by the architect (b). Before issuing a certificate the architect may require reasonable proof from the contractor that earlier certified sums have been paid. In the absence of proof the architect will certify accordingly and the employer may then pay direct and deduct from sums due to the contractor (c).

The architect's written consent is required to a contractor giving an extension of time to a nominated subcontractor whose representations must be passed to the architect (d.i) and when a subcontract is late the architect must certify the period within which the work ought reasonably to have been completed (d.ii).

Where the architect wishes, early final payment may be made to a subcontractor, the contractor to be given satisfactory indemnity by him for latent defects in the subcontract works (e).

There is no privity of contract between employer and subcontractor (f).

A contractor may tender for work the subject of a PC sum and set out in the Appendix, if the architect agrees and it is normally work he carries out (g).

Should a subcontractor fail to complete his contract (eg, because of liquidation), the architect must nominate again and adjust the PC sum so that the extra cost of this does not fall on the main contractor.

Provisional sums may also be expended in this way technically by the creation of a PC sum. Unlike PC sums which are for the nomination of subcontractors and suppliers, provisional sums may extend to work where the cost, extent and exact nature are not fully known at the time of tendering (see section 3.6.5).

The contract sum adjustment is achieved by deducting the PC sum and adding back the final amount due to the nominated subcontractor under the terms of his contract together with percentage additions for contractor's profit as priced. Clause 27 does not apply to statutory undertakers carrying out statutory functions covered by a PC sum (clause 4(3)).

The above references to the pre-1980 Standard Form in respect of nominated subcontractors are confined to this Chapter and should serve to give a full picture of nominated subcontract provisions of the pre-1980 form. Where 1980 Standard Form clause references are given it should be remembered that the 'green' form would not be used with the 1980 main contract. Chapter 3, dealing with the Standard Form, covers the general provisions under the 1980 edition and Chapter 6 discusses the new arrangements from both the main and nominated subcontract angles.

5.3. Non-Nominated Subcontractors

A contractor may decide, either at the tendering stage or later, to ask a specialist firm to carry out work in one of the subtrades such as plastering. This is allowed under clause 17 of the pre-1980 edition and clause 19 of the 1980 edition, but the architect's written consent, which must not be unreasonably withheld, must be sought for subletting. Failure to obtain this could give grounds for determination.

If, at the tendering stage, the contractor decides to sublet some of the work he should send copies of the appropriate section of the bills of quantities to selected firms and, after deciding which offer he prefers, incorporate these prices in his bills before submission of the tender. At this stage the firm chosen should be informed. A firm order would not be placed and a contract made until the main contractor's tender was successful.

If the decision to sublet is taken after the contract has been awarded, then the contractor must be certain that this is in line with his initial pricing.

The terms of subletting may be determined by the main contractor if the subcontractor is prepared to accept them, but the use of the non-nominated subcontract conditions provides an established and mutually acceptable basis for the work. A main contractor who imposed harsh terms on a subcontractor would be acting unwisely since co-operation is obviously important.

The selection of a sole subcontractor by the architect, with his prices incorporated in the main contract bills, was a practice not envisaged in the contract conditions, although provision is now made for this is the 1980 edition of the Standard Form (clause 19) – see section 3.6.7. A 'selected' subcontractor should be clearly distinguished from a nominated subcontractor. For contract purposes he is non-nominated. It should be noted that under the 1980 edition all non-nominated subcontractors are described as 'domestic subcontractors'.

5.4 Main contractor-subcontractor relationships

It is obvious that if contracts are to be efficiently run there must be the maximum understanding and co-operation between the main contractor and his subcontractors. Friction can be avoided by the use of suitable procedures and a clear appreciation of the contract's terms.

5.4.1 PROCEDURES

It is essential that there should be:

(a) careful planning and forethought by all concerned
(b) complete exchange of information
(c) realistic programmes acceptable to all parties.

The following are important considerations:

Nominated subcontractors — Pre-1980 procedure
At the design stage of a building project the architect should decide which work should consist of PC sums and where it should fall in the contractor's programme. Obviously he should nominate in good time and be prepared for renomination to avoid delays although some architects fail to make provision for this. A delay might, under clause 24, give rise to a claim by the main contractor. The subcontractor is deemed to have notice of the terms of the main contract and use of the Standard Form of Tender for Nominated Subcontractors helps to ensure that this is the case. Before a nomination is made the architect should have ascertained that the subcontractor can comply with the main contractor's timetable. As stated previously, a contractor has a reasonable right of objection under clause 27 if a nominated subcontractor's starting date and completion period make the main programme impossible to fulfil. If the main contract is delayed and the programme altered, subcontractors should be informed at once with some indication of revised dates.

The problem of co-ordinating all the subcontract work with the main contract programme calls for special comment. In the initial stages the main contractor should have prepared a detailed programme for each phase of the work showing when each subcontractor should commence and finish his contract. When a nomination is made this information should be available wherever possible so that the subcontractor tenders, the architect nominates, and the main contractor places his order on this basis.

When the subcontract is made, Part II of the Appendix is completed, but this only sets out the period for completing the work or each section of it. As far as starting dates are concerned, clause 8 requires the subcontractor to commence within an agreed time of the main contractor's written order to do so. If there is no agreed time the subcontractor must start within a reasonable time of the order being received. This can be a most unsatisfactory arrangement from the standpoint of the subcontractor. Unless a special provision is included in the contract conditions for starting and finishing dates (and this has its hazards), the starting date would be determined by this notice to commence.

However, in the making of the contract, more precise dates for carrying out the work could have been agreed. This would provide a more efficient basis for the conduct of operations and as a corollary more precise grounds for any claims for delays by either party. At the other extreme, the programme might be the subject of discussion and negotiation even as the contract proceeds. This lack of precision in the contractual provisions themselves can easily lead to claims and counterclaims on delays, hence the need for business-like arrangements. (See section 5.8.2 on claims.)

It will be seen from Chapter 6 that in the new JCT Nominated Subcontract Form extensive and onerous procedures are laid down so as to place nominated subcontract arrangements on the soundest business basis before a contract is made.

Non-nominated subcontractors
Subletting tends to be less business-like than nomination but special enquiry/tender/acceptance procedure, devised to smooth out problems as far as possible, has not been generally used.

Although the period for completing the work should be given in Part II of the Appendix of the non-nominated form, specified dates need not be inserted. Subcontractors should always be given a clear and early indication, if possible at the tendering stage, of when they are likely to be required and whenever the contract is running late. Although the period of notice to commence is also given in Part II, a delayed contract may result in the subcontractor using his resources on other work.

These are some of the main sources of complaint. Difficulties do arise which might have been avoided by care, forethought and an appreciation of the other party's problems. This extends to responsibility for and use of plant and scaffolding, site facilities, including watching and storage, public utilities, health and welfare, safety, attendance, etc. Although these points are generally dealt with in the contract documents, a meeting between contractor and subcontractor would help to clarify the position before work commences on site.

5.4.2 CONTRACT PROVISIONS

How are the responsibilities divided? Where the contract's terms are not fully known and understood, trouble can easily develop. Difficulty and confusion multiply when either party seeks to introduce its own terms at variance with those jointly agreed in the industry. Payment and related problems are singled out first for special comment, and other aspects of the contracts are dealt with in subsequent sections.

5.4.3 PAYMENT

The nominated and non-nominated subcontract conditions provide for cash discount for payments made by the due date – 2½ per cent from nominated subcontractors (clause 11) and 2½ per cent from non-nominated subcontractors unless a different percentage is inserted in Part V of the Appendix of the 'blue' form. This discount, however, is not automatic, and it is incorrect to deduct it, as so often happens, once the due date is past.

The provisions of the contracts are as follows.

Payment — nominated ('green') form (clause 11)

At least seven days' notice must be given to the nominated subcontractor of the contractor's application to the architect for a certificate of payment which will include sums due to the subcontractor. This is to ensure that it includes the up-to-date value of the subcontract works together with variations, fluctuations, loss or expense claims, appropriate materials on and off site (where agreed), all subject to retention.

The amount due to the nominated subcontractor and included in the certificate should be notified to him by both the contractor under clause 11(b) of the subcontract, and by the architect under clause 27(b) of the main contract. This money should be paid by the contractor within 14 days of his *receipt* of the certificate, less 2½ per cent cash discount. This cash discount applies to all payments under the subcontract (eg, fluctuations).

It is important to note the qualification for cash discount and the requirement of the contractor to pay whether his certificate has been honoured or not. When the main contractor is in default then if he fails to pay within seven days after written notice, the subcontractor is entitled to suspend work under clause 11(e). If work is resumed the subcontract period is deemed to be extended for the period of the stoppage. The main contractor, being at fault, would not be entitled to a similar extension under clause 23(g).

The employer may, under clause 27(c) of the main contract, pay the subcontractor direct, deducting the amount from sums due to the main contractor, provided that the architect certifies that the contractor has failed to prove that money due on previous certificates has been paid. It is unlikely that on the insolvency of the main contractor the liquidator will be bound by this arrangement or to pay to the subcontractor sums due to him, still held by the main contractor. (The new provisions in clause 35.13.5.4.4 of the 1980 edition of the Standard Form should be noted in this connection.)

Subcontract payments are subject to the main contract retention percentages and release provisions, but it is specially provided under clause 11(h) that the contractor's interest is fiduciary as a trustee for the subcontractor in respect of his retention (ie, it would appear to be protected in bankruptcy, etc). No special trust fund is created. Cash discount is also allowed on release of retention. Where subcontract work is completed, a special interim payment may be made before the issue of the final certificate under clause 11(f). Under clause 11(g), if the work is completed early in the main contract programme, final payment to the subcontractor may be made in advance of the final certificate, if it is certified by the architect and an indemnity given to the main contractor by the subcontractor for latent defects.

Payment — non-nominated ('blue') form (clause 13)

Payments to non-nominated subcontractors under clause 13 of the contract are not dependent on or related to the issue and honouring of main contract certificates. This means that the main contractor must organise his financial arrangements for payment on due dates irrespective of how he fares on main contract payments.

The first payment is due not later than one month after the commencement of work and, further payments must be made monthly (clause 13(I)). These payments must be made within 14 days when the 2½ per cent or other specified discount may be deducted. They should include the value of work done, materials on site, off site materials, if agreed under main contract, variations, loss or expense, and fluctuations up to seven days previously. Substantiating details must be provided by the subcontractor in support of the value of work, etc claimed (clause 13(2)).

Retention, as set out in Part V of the Appendix, is generally at the rate applying to the main contract, with half released on practical completion of the subcontract works to the satisfaction of the contractor and architect, and the other half either on the release of the retention under the main contract or earlier if specified, subject to indemnities (clause 13(3)).

As in the nominated form, failure to pay for seven days after written notice gives grounds for suspending work under the subcontract until payment is made.

Factors influencing payments

In the past few years three major considerations have been the subject of contractual provisions which are of some complexity. These are briefly as follows:

(1) The set-off position has been dealt with in lengthy provisions involving references to an adjudicator, etc ('green' form clauses 13A and B; and 'blue' form clauses 15 and 16) (introduced in February 1976).
(2) VAT is covered by 'green' form clauses 10A and B and 'blue' form clauses 12A and B allowing for the alternative of self-billing (introduced in November 1975).
(3) The new subcontractor tax deduction scheme, which came into effect on 6th April 1977, made amendments to the subcontract forms desirable ('green' form clauses 10C and D; 'blue' form clauses 12C and D).

Since December 1978 there have been no contractual provisions on incomes policy. A section in Chapter 4 gives information on various requirements from 1973 to 1978 including special clauses 11A ('green') and 13A ('blue') and the short-lived special subcontract provisions necessitated by Supplementary Government Conditions 185, 186 and 187 operating from March to December 1978.

Detailed explanations are given below of the subcontract provisions on set-off, VAT and the tax deduction scheme.

5.5 Set-off

Court decisions on set-off in recent years provided fruitful ground for legal argument. Since it is clear that the courts will normally enforce the contractual arrangements for set-off made between the parties, the position in building subcontracts has been clarified by new provisions introduced in February 1976. These provisions on set-off are contained in clauses 13A and 13B of the 'green' nominated form and clauses 15 and 16 of the 'blue' non-nominated form. They are almost identical. (See also Chapter 6 for provisions in the JCT Nominated Subcontract Form, NSC/4 and 4a, clauses 23 and 24.)

There are two separate aspects of the subject – the right of set-off and the course open to the subcontractor if he does not agree with the amounts deducted from payments due (ie, the appointment of an adjudicator).

5.5.1 RIGHT OF SET–OFF

Clause 13A(1) of the 'green' form and 15(1) of the 'blue' form entitle the contractor to deduct from sums due (including retention), amounts agreed by the subcontractor as due or finally awarded in arbitration or litigation in respect of that particular subcontract. The cash discount is calculated on the net amount (ie, less set-off.) Where there is a loss and or expense claim or one arising from a breach of, or failure to observe the subcontract provisions, and the amounts are not agreed by the subcontractor, the following conditions must be met before there is a set-off entitlement:

(i) The amount of any set-off must be quantified in detail with reasonable accuracy.
(ii) At least 17 days before the payment is due, the proposed set-off must be notified in writing by the contractor to the subcontractor, with details and grounds for the claim.

In the nominated subcontract form there is the additional condition that if delay is claimed there must first be a certificate from the architect under clause 8(a), the duplicate copy having been sent to the subcontractor.

In both forms there are three provisos:

(i) The amount claimed and set-off is without prejudice to the rights of either party to seek to vary it in subsequent negotiation, arbitration or litigation.
(ii) Claims and counterclaims under the adjudication procedure are not binding in any subsequent arbitration pleadings and may be amended.
(iii) The rights of set-off are those set out in the contract conditions and no other rights shall be implied.

5.5.2 APPOINTMENT OF AN ADJUDICATOR

If the subcontractor wishes as a matter of urgency to dispute the contractor's right of set-off there is relatively speedy procedure involving notice of arbitration and reference to the adjudicator named in the Appendix to the subcontract. The procedure is rather involved but it can be summarised as follows (all statements must be sent by registered post or recorded delivery).

(a) If the subcontractor does not agree the amount proposed for set-off in the contractor's notice (see above) he may within 14 days of its receipt send a written statement to the contractor setting out his reasons for not agreeing and giving quantified and accurate details of any counterclaim.
(b) The subcontractor must at the same time give notice of arbitration to the contractor and request action by the adjudicator named in the contract (informing the contractor at once). He must send a copy of the contractor's notice and his statement and any counterclaim to the adjudicator.
(c) On receiving the subcontractor's statement the contractor may within 14 days of receipt send written particulars of his defence to the counterclaim to the adjudicator (with a copy to the subcontractor).
(d) Within seven days of the receipt of the contractor's defence, or on the expiry of 14 days from the contractor's receipt of the subcontractor's statement, whichever is earlier, the adjudicator without requesting further statements (except to clarify any point) and without hearing either party must reach a decision in his absolute discretion and without giving reasons, on the basis of the whole or part of the amount set-off: (i) to be retained by the contractor; (ii) to be paid to the subcontractor; or (iii) pending arbitration deposited by the contractor with the trustee-stakeholder named in Part XIII of the

Appendix to the 'green' form (Part XIV – 'blue' form).

(e) The decision which the adjudicator considers fair, reasonable and necessary must be notified at once to the parties in writing and must be complied with immediately. It must deal with the whole amount set off and is binding until the issue is resolved by agreement between the parties or before an arbitrator or in the courts. The amount a contractor may be obliged to pay over to the trustee-stakeholder must not exceed the amount certified as due to the subcontractor in respect of which the right of set-off is being exercised.

(f) If either party is dissatisfied with the adjudicator's decision, arbitration, of which notice has already been given, can follow promptly. If an arbitrator is appointed he may, before his final award, vary or cancel the adjudicator's decision on the application of either party.

(g) Any action taken under these provisions will not prejudice future action in respect of further sums.

The following points should be noted:

(1) *Adjudicator*

An up-to-date list of those prepared to act (almost entirely quantity surveyors) is available from the NFBTE. Prior consent to appointment is not required; any nominee unable or unwilling to act must appoint someone in his place. Adjudicators appointed must have no interest in any contract of either main contractor or subcontractor. The adjudicator's fee must be paid in the first instance by the subcontractor within 28 days of the decision. If the matter goes to arbitration the final award will settle responsibility for the adjudicator's fee and stakeholder's charges.

(2) *Stakeholder*

The trustee-stakeholder, whose agreement should first be obtained, is named in the Appendix to the contract, (eg, a main clearing bank or alternatively a solicitor might be prepared to act). The trustee-stakeholder will hold the sum determined by the adjudicator, on deposit account in his name, to which interest will be credited. From this amount the trustee-stakeholder will be entitled to deduct his reasonable and proper charges. The amounts received must be held and not disposed of until directed by the arbitrator, or in the absence of such a direction, as the contractor and subcontractor jointly determine in writing. It is the duty of the subcontractor to notify the trustee-stakeholder of the name and address of the adjudicator and arbitrator.

5.6 VAT

There are alternative clauses dealing with VAT – one where the subcontractor provides the tax receipts ('green' clause 10A; 'blue' clause 12A) and the alternative covering self-billing arrangements ('green' clause 10B; 'blue' clause 12B).

In alternative 'A' both forms of contract make it clear that the contract sum is VAT exclusive, the tax chargeable to be additional to the subcontract sum ('green' clause 10A.2; 'blue' clause 12A.2). Prior to 1 June 1975 all subcontract work was positively rated for VAT. From that date subcontract new work (or alterations or demolition) was zero rated as in main contract work. To enable the main contractor to add the correct sum for VAT in any payments, the subcontractor is now required, not later than seven days before the date payment is due, to give the main contractor a provisional written assessment of zero rated supplies and those positively rated, with reasons and the rate applying ('green' clause 10A.4a.ii; and 'blue' clause 12A.4a.ii).

There is a safeguarding provision on tax receipts. To the amounts due under 'green' clause 11 and 'blue' clause 13 the main contractor must calculate and add VAT and the subcontractor on receiving payment must issue a tax receipt. If within 21 days this has not been received, further payments, may, on written notice, be withheld by the main contractor until a tax receipt is received ('green' clause 10A.5 and 'blue' clause 12A.5). It is made clear that if cash discounts wrongly deducted are refunded to the subcontractor, this procedure does not apply.

In November 1975 clause 10B was introduced in the 'green' form and clause 12B in the 'blue' form. These are alternative clauses to 10A and 12A which provide for the subcontractor preparing authenticated receipts for VAT included in payments by the contractor. The alternative arrangement is self-billing, ie, a system approved by Customs and Excise, and agreed to by the subcontractor, where the contractor prepares his own input credit documents. The provisions as in 10A (or 12A) apply, with the following special procedures. Where a contractor makes a payment to a subcontractor including VAT he is required to issue a document approved by Customs and Excise giving only the date of dispatch. When the subcontractor receives both the payment and the document, the date of receipt (ie, the tax point) should be inserted. The date of receipt should not however be inserted until payment is made and if it is for a lesser amount a reconciliation is necessary. While this operates the subcontractor must not issue authenticated receipts.

This arrangement continues as long as it is approved by Customs and Excise and agreed between the parties. If approval is withdrawn or the subcontractor ends the arrangement and notifies Customs and Excise and the contractor, clause 10A (or 12A) thereafter applies.

Note: The 1980 JCT nominated subcontract provisions on VAT are in clauses 19A and 19B of NSC/4.

5.7 Construction Industry – Tax Deduction Scheme

The new scheme under the Finance (No 2) Act 1975 which came into operation on 6 April 1977 is outlined in section 3.8.5.

The previous (1972) scheme did not affect main contracts since the employer was not a contractor for the purposes of the scheme.

The provisions in the pre-1980 Standard Form of Contract (clause 30B) which were new, were necessary because of the very much wider definition of a 'contractor' under the new scheme which includes public authorities, etc. Thus the main contractor becomes a 'subcontractor' under the scheme where the employer by definition is a 'contractor'.

The provisions in clause 30B (clause 31 – 1980 edition) are broadly similar to those introduced into the nominated 'green' form (clauses 10C and 10D) and the non-nominated 'blue' form (clauses 12C and 12D). (See section 3.8.5 for main contract provisions.)

Under the 1977 scheme all subcontractors carrying out 'construction operations' under the 'blue' or 'green' forms are 'subcontractors' under these tax deduction provisions. The earlier 1972 scheme gave rise to an optional provision in the 'green' and 'blue' forms to be deleted if so desired: thus in the recital it was stated that the subcontractor was exempt from the provisions of the Finance Act 1971 (Sections 29-31) and 'green' form clause 20(b) ('blue' form clause 21(2)) included a provision that if the subcontractor ceased to be exempt the contractor might determine the contract. This no longer appears in the current forms.

The resultant amendments to the 'green' and 'blue' forms cover the alternative situations: clauses 10C and 12C where the subcontractor has a current valid tax certificate and clauses 10D and 12D where he has not. A changeover from one set of provisions to the other is envisaged where, during the currency of a subcontract, a subcontractor obtains or ceases to hold a current valid tax certificate.

The new clauses are designed (as in the Standard Form – see Chapter 3) to co-ordinate provisions on payments and provide remedies and protection for the parties. Since under the Inland Revenue scheme there is no comprehensive timetable (eg, for producing tax certificates) one is introduced into both main and subcontracts to assist in their orderly administration. If payments are affected because the subcontractor fails to comply with the contract's requirements, the responsibility will be his.

The amendments were dated November 1976, and their use prior to April 1977 was recommended as the scheme related to all payments on and after 6 April 1977 (ie, including contracts then current).

5.7.1 CONTRACT PROVISIONS

The provisions in the 'green' and 'blue' forms are virtually identical, the outline below applying equally to both forms. ('The 'blue' form provisions are found in clauses 12C and 12D).

In the nominated 'green' form of subcontract the third recital is amended in the light of the new subcontractors' tax deduction scheme and sets out the following alternatives:

(i) whether or not the subcontractor and the main contractor have current valid tax certificates
(ii) whether or not the employer is a contractor within the meaning of the Act.

The alternative clauses are 10C, where the subcontractor has a current valid tax certificate at the date of the subcontract, and 10D where he has not (to tie in with recital, as above).

All this information is essential from the main contractor/subcontractor standpoint. It is also vital for the subcontractor to know whether the employer is the 'contractor', eg, a local authority (this information now appears in the Recitals and Appendix of the 1980 edition of the Standard Form) and if so, whether the main contractor has a valid tax certificate. Payments net to the main contractor would have to exclude from the calculation the direct cost of materials not only to the main contractor but also to subcontractors, where payments to them are included. If the main contractor does not have a current valid tax certificate and payments to him are made net, the subcontractor with a valid certificate would insist on payments to him without deduction of tax.

The provisions of 10C ensure that the scheme's requirements are fully observed in keeping with a prescribed timetable. The subclauses provide as follows the references in brackets being to subclause numbers:

(a) Payments are to be made without deduction where a current valid tax certificate (or the certifying document for 'C' certificates) has been either (a) produced not less than 14 days before the first payment is due, the contractor notifying the subcontractor in writing within seven days of its production that he is satisfied that the regulations have been complied with (subclause 1.(a); or (b) previously produced to the contractor who has confirmed in writing his satisfaction (1(b)). If the contractor is not satisfied under the regulations, the subcontractor must be so informed in writing within seven days of the production of the evidence (1(a)).

(b) The subcontractor must at once notify the contractor in writing of any change in the nominated bank account, under the certifying document method (1(c)). Similarly the contractor must notify the subcontractor of any change in his own certificate position or that of the employer as set out in recital 3.(2)(b).

(c) The subcontractor with a 714 I or 714 P certificate (ie, not having a 714 C company certificate) must when any payment is made immediately issue a 715 receipt (3(a)) which the contractor must immediately pass on to the Inland Revenue (3(b)).

(d) If a subcontractor no longer has a valid certificate (when payments would have to be made net), before any payment is due the contractor must immediately notify the subcontractor in writing accordingly and require him to state not later than seven days before the due date, the proportion of the amount due, attributable to the direct cost of materials. From this amount there is no tax deduction (4(a)). The contractor is to be indemnified against loss where the statement is incorrect (4(b)). If the subcontractor does not provide this information the contractor may make a fair estimate of the direct cost of materials in calculating the tax deduction (4(c)).

(e) Errors or omissions are to be corrected and adjusted by the contractor subject to any statutory obligations (5).

In clause 10D covering payments made subject to deduction of tax to subcontractors not possessing a valid certificate, the procedure as in (d) and (e) above apply (subclauses 1 and 2). If the subcontractor subsequently obtains a valid certificate up to and including the date when the last payment is due, he must inform the contractor in writing at once when 10C will apply from the date of notification (5). It is the duty of the contractor to inform the subcontractor immediately of any change in his certificate position or that of the employer as set out in recital 3 (3).

Tax is deducted from the net payment to a subcontractor (ie, less cash discount and set-off, if any).

Both clauses stipulate that these provisions (which have legal sanction) will prevail if in conflict with any other clause. The contract's arbitration provisions apply to any dispute unless there is a statutory alternative laid down under tax legislation.

(Note: The 1980 JCT nominated subcontract provisions on the tax deduction scheme are in clauses 20A and 20B of NSC/4).

5.8 Subcontract forms – similarity in provisions

The most effective way of explaining the subcontract conditions is to set out those provisions which are similar in the nominated and non-nominated forms and then to consider the specific terms of each. In this section, to distinguish the forms concerned, clause references are given as follows: g6: b4. This means 'green' (nominated subcontract) form, clause 6, and 'blue' (non-nominated subcontract) form, clause 4. The main contract references are to the pre-1980 form, but where of assistance the 1980 clause references are given in brackets.

The basic principle in devising these forms is that in respect of the subcontract works the subcontractor should broadly accept the responsibilities of the main contractor under the main contract.

5.8.1 GENERAL CONTRACTUAL CONDITIONS

(a) The subcontractor is deemed to have notice of and the opportunity of inspecting the provisions of the main contract other than prices (g1: b1).

(b) The non-nominated subcontractor must carry out the work to the reasonable satisfaction of the main contractor and the architect (b2). To accord with clause 1(1) (now 2.1) of the pre-1980 Standard Form the nominated subcontractor must carry out the work in compliance with the subcontract documents, to the main contractor's requirements, using materials and workmanship of the quality and standards specified which, where required, must be to the reasonable satisfaction of the architect (g2). The architect's written instructions issued in writing by the main contractor to the subcontractor must be complied with at once (g7(1): b8(1) and (2) – See g7(2) on the right to challenge). If there is failure to comply, after seven days' notice in writing requiring compliance, the contractor may have the work done at the subcontractor's expense (b8(4)).

(c) The provisions of the main contract, insofar as they relate to the subcontract, must be fully complied with, subject to the express terms of the subcontract (g3: b3). The main contractor is to obtain for subcontractors appropriate rights and benefits of the main contract (g12: b14). This provides, for example, for loss or expense claims under clause 24 (now 26) where progress of the main contract is disturbed (g8(c): b10(1)): extensions of time under clause 23 (now 25) (g8(b): b9(3)); defects where made good other than at the main contractor's expense (g9(a): b11(3)).

(d) Defects must be made good within a reasonable time in accordance with a written instruction of the architect (g9(a): b11(2)) or the direction of the contractor (b11(2) only) (see section 5.10 for grounds for determination.) If in a nominated subcontract the work of one party is defective and affects the work of the other the expense incurred must be met by the party responsible (g9(b) and (c)).

5.8.2 RESPONSIBILITIES OF PARTIES TO EACH OTHER

(a) If attributable to the sub-contractor, his servants or agents, the contractor must be indemnified by the subcontractor against breach, etc, of any of the main contract conditions; any act or omission rendering the contractor liable to the employer under the main contract, or any claim, etc, due to negligence or breach of duty on the part of the subcontractor including the wrongful use of scaffolding, (g3: b3).

(b) There are appropriate provisions covering liability, etc through injury to subcontractor's employees and to persons or property and the responsibilities for insuring (g3(iv) and 4:b4)

(c) Loss or damage by fire, etc to the subcontract works and materials properly on site for incorporation is at the sole risk of the contractor (g5(a): b5(1)), the risk to be insured and the amount of the loss (including fees) to be paid to the subcontractor (g5(c): b5(3)).

(d) Any loss or damage, other than fire, etc, but including theft or vandalism to subcontractor's unfixed goods and materials shall be the responsibility of the subcontractor until 'fully finally and properly incorporated into the works' unless damage is due to the negligence, etc of the main contractor, any other subcontractor or the employer, their servants or agents. This applies to part of the works even though the whole of the subcontract works may not have been completed (g5(b): b5(2)). Responsibility for loss or damage to unfixed goods and materials once incorporated passes to the contractor, together with the whole of the subcontract works when completed, except of course where any loss or damage was caused by the subcontractor his servants or agents. Unless caused by the main contractor's negligence, the subcontractor is responsible for loss or damage to his plant and tools (and materials not properly on site for incorporation) and should insure accordingly (g19: b7).

(e) Evidence of the required insurances must be produced to the other party when required (g6:b6).

(f) In broad terms claims can arise on both contracts as follows, in relation to programmes and progress:

By main contractor against subcontractor

(i) For failure to complete on time (with authorised extensions) (g8(a):b9(3)). Damages may be predetermined (eg, same figure as in the main contract) or more frequently and realistically, left open to be agreed between the parties when the extent of the loss suffered can be more accurately assessed. It is a condition prior to a claim that the architect must certify the period within which the work ought reasonably to have been completed (g8(a)) and the contractor must notify the loss or damage as soon as possible to the subcontractor.

(ii) For disturbance of regular progress of the main contract because of any act omission or default of the subcontractor his servants or agents (prompt written notice to be given) (g8(c)(iii): b10(3)). The amount agreed is deducted from the contract sum.
Note: Delays by a nominated subcontractor (or supplier) which the main contractor has taken all practicable steps to avoid or reduce give grounds for an extension of time under clause 23(g) (now 25.4.7) of the Standard Form.

By subcontractor against main contractor

For materially affecting the regular progress of the subcontract works through any act, omission or default of the contractor his servants or agents or any other subcontractor (prompt written notice to be given) (g8(c)(ii): b10(2)). The amount of any direct loss or expense agreed is added to the contract sum.

A nominated subcontractor may also be delayed for two other main reasons – those for which the main contractor could under the main contract make a claim for loss and or expense under clause 24(now 26) (g.8c(i)) and those affecting his work for which an extension of time could be sought (g.8b) claims being dependent on the reasons for the delay.

A controversial point, and one on which the contract conditions are not explicit, is whether if one party is delayed without default and is given an extension of time, the other party can claim damages for delay in respect of that period. Equitably one feels he should not be able to do so (eg, a main contractor is granted an extension of time for exceptionally inclement weather and the start of a subcontractor's work is delayed. If this had happened to the subcontractor he also would have qualified for an extension and a claim on the main contractor in these circumstances would not appear to be reasonable).

Note: Particularly on required insurance cover, the provisions of the contracts should not be altered in any way to shift responsibility from one party to another.

5.9 Other provisions

5.9.1 FLUCTUATIONS

The provisions for subcontract fluctuations are very similar to those in clause 31 (now clauses 37 to 40) of the main contract, fully explained in Chapter 4.

In the nominated ('green') subcontract form the appropriate clauses are 23A or B, C, D and E for the traditional method and clause 23F for the price adjustment formula. The clauses in the non-nominated ('blue') form are 25A or B; with C, D and E; and F.

One relatively minor difference is that nominated subcontract 'traditional' fluctuations are subject to retention (g11(b)) whereas in the main form they are not.

If the subcontractor sublets he must incorporate in that subcontract 'traditional' fluctuations provisions similar to those in his subcontract (g23C(1): b25C(1)). This parallels clause 31C(1) of the pre-1980 main form and provides the basis for fluctuations recovery in respect of this sublet work.

The interrelationship of main and subcontract fluctuations provisions has been the subject of the following guidelines agreed between the industry and local authority associations which are recommended for local authority contracts (See section 2.4.5).

(i) If a main contract of two or three years' duration is let on the traditional method, nominated subcontracts may be let on the formula basis if the subcontractors prefer it and the authority so decides.

(ii) Where the main contract is on a fluctuations basis, nominated subcontracts should be let on this basis if the subcontract work is not properly pre-planned, if it is to last more than 12 months, or if the starting and finishing dates are not certain at the time of tender.

(iii) In non-nominated subcontracts the arrangements must be agreed between the parties but payment to the subcontractor should only be on the main contract basis (ie, the subcontract fluctuations provisions should preferably match, but certainly be no more favourable than, those of the main contract).

It is, of course, open to the parties under the private work form to agree such arrangements as are mutually acceptable but these points can usefully be borne in mind.

Several aspects of the application of the price adjustment formula to subcontracts call for special consideration. Special subcontract formula rules for both nominated and non-nominated subcontracts were issued on 3 May 1975 under Series 1 and on 4 April 1977 under Series 2, and are an integral part of clause 23F and 25F. These are in two parts: Part 1 sets out the work category method which is used where the subcontract work is of a nature which can be covered by one or more of the 34 (Series 1) or 48 (Series 2) work categories (see Chapter 4). Work groups are not recognised in the subcontract provisions. Part II provides specialist formulae for engineering installations – electrical; heating, ventilating and air conditioning; lifts; structural steel; and in Series 2 catering equipment installations. These require careful study. All information on the formula to be used must be given in the tendering documents and details entered in Parts IX, X, XI and XII of the Appendix ('green' form) or Part X, XI, XII and XIII of the Appendix ('blue' form).

The general principles are set out in detail in section 4.4 where it will be noted in section 4.4.1 that on the introduction of the 1980 JCT Nominated Form, the Formula rules have been revised. The following considerations should be noted in using the formula adjustment in subcontracts.

The non-adjustable element percentage in the main contract (local authorities edition) must apply to the subcontract, but if clause 31F does not apply to the main contract the non-adjustable element percentage (not exceeding 10) for the nominated subcontract must be inserted in the Appendix, Part X.

The principle of 'freezing' the indices under main contract clause 31F.7 (now 39.7) (see section 4.5) applies to subcontract formula adjustments, ie, at the end of the subcontract period set out in Part II of the Appendix as extended under clause 8(b) ('green') or clause 9(3)('blue'). In the nominated form this is subject to the architect having fully dealt with extensions under clause 8(b).

References to the main form clause 23, in clause 8(b)(ii) ('green') and clause 9(3)(a) ('blue') are to that clause unamended with 23(j)(i) and (ii) operating (ie, there is no restriction of the grounds for extension of time).

Each contract is treated separately. 'Freezing' will only apply to a subcontract when it is late. If in the main contract the indices are 'frozen' because the main contractor is late, provided the subcontractor is on time, he would not similarly suffer. If the main contract is on clause 31A, specific subcontract valuations would be needed and 'freezing' could take place in a subcontract where 23F was operating. (See note in section 5.13 below on revisions of the 'blue' form).

There are several differences between clause 23F ('green') and 25F ('blue'). These are largely due to differences in procedures and relationships – valuations, payments, functions of architect and quantity surveyor, etc.

These points are made briefly and should be considered in relation to Chapter 4 in which the price adjustment formula is fully explained.

Special attention is drawn to the note under section 5.13 below on necessary revisions to the non-nominated ('blue') form when used for work sublet under clause 19 of the 1980 edition of the Standard Form.

5.9.2 MISCELLANEOUS MATTERS

Access to works must be available for contractor and architect (g14: b17);

Assigning and subletting must be with the consent of contractor (b18), and the consent of both contractor and architect (g15).

Provisions are made for the contractor where required in the main contract, to supply at his own cost, lighting, watching and attendance for nominated subcontractors. The nominated subcontractor must provide at his own expense workshops, sheds, etc, the contractor to give reasonable facilitites for erection (g16). Certain differences should be noted on scaffolding: eg, non-nominated subcontractors are to provide and erect scaffolding for work 11 ft high or under (b19(2)); nominated subcontractors are entitled to use main contractor's standing scaffolding where available (g17). The contractor must supply the non-nominated subcontractor with storage facilities for materials.

There are provisions for fair wages (g22:b24).

5.9.3. ARBITRATION

Arbitration has a similar procedure in both subcontract forms. The arbitrator, unless jointly agreed, is to be appointed on request of either party by the President of the Royal Institution of Chartered Surveyors (g24: b27). Except on the questions of certificates, extension of time or an issue arising from the decision of an adjudicator under clause 13B(g) or 16(b), arbitration cannot take place until the completion of the main contract unless otherwise agreed in writing by the architect, main contractor and subcontractor. An arbitrator's decision on the main contract is binding on the subcontractor who should be fully informed on any developments so that he can make any necessary representations. If an issue is substantially the same as one under the main contract the parties agree that the dispute or difference be referred to the arbitrator under the main contract, whose decision will be final and binding on all parties (see section 3.10 for a full note on arbitration).

5.10 Determination of Contracts

Determination of the subcontractor's employment by the main contractor (g20: b21) may be effected for suspension of work without reasonable cause, failing to proceed diligently with the work, or refusing after due notice to remove defective work and material. If the default continues for 10 days after notice by registered post, notice of determination may then be given also by registered post. Although in a nominated subcontract the prior consent of the architect is not apparently specifically required to determine, since he would be asked to renominate, full consultation and approval are desirable with an assurance that no additional costs will fall on the contractor. Where bankruptcy or liquidation of the subcontractor occurs, immediate written notice of determination should be given, provided the main contractor is not also in breach.

Only in the non-nominated form (b.22) may the subcontractor determine the contract. (This right now appears in NSC/4 clause 30) There is provision in both subcontracts for the valuation of work at determination (ie, value of work not paid for and unfixed goods and materials the property of the employer) and claims by the main contractor for loss incurred or damage suffered, with a right of deduction or set-off.

If the main contract is determined, then the subcontract is also determined, the subcontractor's payment entitlements to include the value of work completed and executed, unfixed materials which have become the property of the employer, materials ordered and paid for, cost of removing sheds, plant, etc but not, apparently, any other loss suffered (g21 : b23).

5.11 Nominated subcontract form – special features

The nominated subcontractor is much more affected by the decisions and instructions of the architect than the non-nominated subcontractor.

For example, a right of reference to arbitration under the main contract can be exercised by the nominated subcontractor through the main contractor, that is, by the use of his name, subject to such indemnity and security as the main contractor may require in respect of:

(a) a challenge of the authority of the architect under the contract to issue a particular instruction (Standard Form clause 2) (clause 7(2)).
(b) the failure of an architect to agree to an extension of time on subcontract work which is considered reasonable (clause 8(b))
(c) the failure to certify an amount considered due (clause 11 (d)).

Under clause 7(1) the subcontractor must comply with variations, etc issued by the architect through the main contractor. The valuation is determined by the quantity surveyor under the main contract, and the subcontractor is given the opportunity of being present at any measurement or remeasurement. The rules for valuation set out in clause 11 of the main contract are generally to apply (clause 10(b)) amounts ascertained to be included in the next certificate (clause 10(c)). The architect must also consider a written application from the contractor on behalf of a subcontractor for a claim which would normally fall under clause 11(6) of the Standard Form (clause 10(d)).

In regard to completion, the contractor is not able to proceed with a claim for loss or damage for delay in completion by the subcontractor unless the architect certifies in writing to the contractor (with a copy to the subcontractor) that the work ought reasonably to have been completed within the period specified or any extended period (clause 8(a)).

Extensions of time for the reasons set out in clause 8(b) such as variations under clause 7(1); the default of the contractor or any of his subcontractors; or for reasons in clause 23 of the main contract under which the main contractor can claim an extension of time, can only be given by the contractor on the written consent of the architect.

5.12 The employer and the nominated subcontractor

The employer is not a party to the nominated subcontract. Therefore he has no right of recourse against the nominated subcontractor if, for example, he is late on the subcontract and the main contractor claims an extension of time. The employer has to put up with the delay and, because the contract period is properly extended, there can be no claim for liquidated and ascertained damages.

At the same time, there may be a design element in a specialist subcontractor's work and, since the Standard Form of Contract deals in general with construction and not design, a failure in design might make it necessary for privity of contract to be established directly between the employer and the subcontractor, if the former is to have any remedy.

To meet this, the RIBA published a form of agreement (current edition 1973 and superseding the 1969 Warranty Form) to be entered into between the employer and a subcontractor nominated under clause 27 of the Standard Form to be effective when the main contractor's order has been accepted. The main contractor is in no way a party to this agreement. At the enquiry stage architects should state if this agreement is to be entered into. The 1980 JCT agreement is NSC/2 or 2a (see section 6.2.1.)

In the agreement the subcontractor warrants to the employer that all reasonable skill and care has been and will be exercised in the design, the selection of goods and materials, and the satisfaction of a performance specification where these are part of the subcontract terms. If any of these do not apply they should be deleted and initialled.

Except for extensions of time allowed under clause 8(b)(i) and (ii) of the Standard Form of subcontract, the subcontractor warrants that he will supply such information as the architect or contractor may reasonably require and so carry out the work that the main contractor has no entitlement to an extension of time under clause 23(g). That is, the nominated subcontractor makes himself responsible to the employer, once a subcontract has been entered into, for loss arising through failure to complete on time for reasons attributable to him. These warranties must not be limited by the subcontractor's tender.

The employer, on the other hand, undertakes without prejudice to the operation of clause 27(c) of the main contract.

(i) to pay the subcontractor direct, including VAT as appropriate, where proof of earlier payment has not been given by the main contractor and the architect has certified accordingly (this is more positive than clause 27(c) of the Standard Form which says that the employer 'may pay')
(ii) where sums are in dispute, to regard payment by the contractor to a stakeholder, pending settlement, as proof of discharge to the extent of that payment.

There is a proviso that after practical completion or in the event of determination of the main contractor's employment by the employer under clause 25(2) (main contractor's bankruptcy or liquidation), only monies in the hands of the employer or under clause 30 will be paid over. This is subject to funds needed to remedy defects, to meet loss or expense on determination or to make payments to other nominated subcontractors under clause 27(c).

It is important to emphasise once again that there are certain major differences in provisions and procedures between the 'green' form and the new JCT Nominated Subcontract Form. The comparative table in Chapter 6 should be carefully studied.

5.13 Non-nominated subcontract form – special features

Although variations may arise from a written instruction by the architect, passed on by the main contractor, or from a direction from the main contractor himself (clause 8), the valuation of variations is a matter between the main contractor and his subcontractor. The rules for valuation in clause 11 (now clause 13) of the main contract generally apply. If daywork rates are set out in Part IV of the Appendix these will apply to extra work which cannot properly be measured or valued; if not, the specified percentage additions to prime cost as defined will operate.

When, in his opinion, the subcontract works are practically completed, the subcontractor must notify the contractor in writing. This date stands unless the main contractor dissents within 14 days in writing giving reasons. If he does, a date must then be mutually agreed (clause 11). From this date responsibility for damage to the subcontract works passes to the main contractor (unless due to damage by the subcontractor) and half the retention is released (clause 13(3)).

Extensions of time must be agreed between the main contractor and his subcontractor for the reasons set out in clause 9(3) which are similar to those in clause 8(b) of the nominated form, where however the contractor must obtain the written consent of the architect to extensions given.

Under clause 22 of the non-nominated form the subcontractor is entitled to determine his contract by registered post or recorded delivery where the main contractor is in default because, without reasonable cause, he wholly suspends the main contract works before completion, fails to proceed with them so that the reasonable progress of his subcontractor is seriously affected, or fails to make payment in accordance with the contract. This default must have continued for ten days after notice by registered post or recorded delivery specifying the default. The resultant position and payments due are set out in clause 22(2).

Note: Since the Non-nominated Subcontract Form will continue to be used where work is sublet to domestic subcontractors under clause 19 of the 1980 edition of the Standard Form, some revision of its terms will be necessary and these are likely to be available in time for use with the 1980 Main Form. One revision in particular will be to clause 25A etc (fluctuations) to bring it into line with clauses 37 to 40 of the main form (see section 4.2) and to restrict recovery of fluctuations under clauses 25A, B or C where the subcontract is late, in terms similar to clauses 38.4.7 and 39.5.7 of the main form (see section 4.2.5).

A short form of the Non-nominated Subcontract was produced in 1979. Brevity is achieved by retaining the Articles of Agreement, Preambles and Appendix and incorporating the full provisions by reference. The provisions themselves are no shorter nor less involved!

5.14 Non-nominated Subcontract – special procedures

In 1971 the NFBTE in association with FASS and CASEC produced a series of three forms to clarify essential tendering and contracting points in an orderly manner.

These were the Enquiry, the Standard Form of Tender and the Standard Form of Acceptance. Being elaborate they were not widely used but for the record they are reviewed below.

STANDARD FORM OF ENQUIRY

This is an enquiry from a contractor to selected subcontractors giving details of the main contract and inviting tenders by a given date for subcontract work as fully described, including timetable and site requirements, either on priced bills or schedules of rates as appropriate, the standard conditions of non-nominated subcontract to apply.

STANDARD FORM OF TENDER

This is the subcontractor's offer in response to the Enquiry, bills of quantities, where appropriate, being part of the tender, together with daywork rates and provisions on fluctuations where they apply. If the offer is on a different basis to the Enquiry on the timetable, site requirements, etc, this must be shown. The form is in duplicate so that the subcontractor retains a copy.

STANDARD FORM OF ACCEPTANCE

This is the main contractor's acceptance to which is annexed in the subcontractor's copy, the Standard Conditions for Non-Nominated Subcontract works. It is prepared in duplicate, both parties signing. There is no appendix to the conditions since the relevant matters have already been covered by the Enquiry and Tender.

OTHER PROCEDURES

There is a simple Form of Enquiry for subcontract works, published by the NFBTE. This gives brief details of the contract and invites subcontractors to tender on the information given which may include drawings and a specification. There are ten simple conditions which, on the acceptance of the tender, become binding on the parties. A section for Special Conditions is provided for details relating to the particular subcontract. This is designed for small sublet work where the full-scale form is excessively detailed. The NFBTE is at present considering a Code of Procedure for Letting and Management of Non-nominated Subcontract Works.

6

NOMINATED SUBCONTRACT CONDITIONS – THE JCT FORM

6.1 General review of new arrangements

It was as far back as June 1966 that the Joint Contracts Tribunal received the approval of its constituent bodies to include in its membership the specialist subcontractor organisations, FASS and CASEC, and to extend its responsibilities to the provision of a standard form of nominated subcontract and to set down procedures leading to nomination. After considerable discussion and revision, drafts were published in 1974, but further consideration was necessary to reconcile the various points of view expressed. It was not until January 1980 that the new form was published. This supersedes the Standard Form of Subcontract (the 'green' form) issued under the sanction of the NFBTE and FASS and approved by CASEC the publication of which will cease. The 'green' form must not be used with the 1980 edition of the Standard Form but its provisions will be significant for some time yet, since many contracts entered into under those conditions will continue to operate after that date. Chapter 5 therefore deals fully with the procedures and provisions of the 'green' (nominated) and also the 'blue' (non-nominated) forms of subcontract and should be read in conjunction with this chapter. Since the JCT Nominated Subcontract Form takes the place of the 'green' form with which the industry is well acquainted, a comparison of the provisions is set out later in this chapter.

For subcontracts entered into under the 1980 edition of the Standard Form the subcontract relationships can be summarised as follows, the (main) Standard Form of Contract (1980 edition) having been amended to provide for the new arrangements.

6.1.1 NOMINATED (JCT FORM)

Nominations will be subject to the terms of clause 35 of the Standard Form of Building Contract (1980 edition) The basic method detailed below, applies unless otherwise stated in the bills or specification or in any instruction on a variation or the expenditure of a provisional sum.

The new nominated subcontract documents are as follows:

(i) JCT Standard Form of Nominated Subcontract Tender and Agreement (*Tender NSC/1*)

(ii) JCT Standard Form of Employer/Nominated Subcontractor Agreement (*Agreement NSC/2*)
(iii) JCT Standard Form for Nomination of a Subcontractor where Tender NSC/1 has been used (*Nomination NSC/3*)
(iv) JCT Standard Form of Subcontract for Subcontractors who have tendered on NSC/1 and executed Agreement NSC/2 and been nominated by Nomination NSC/3 (NSC/4)

If a design service is provided by a specialist nominated subcontractor this is covered in NSC/2 to which the employer and nominated subcontractor are parties but not the architect or the main contractor (see 6.2.1 below – Agreement NSC/2).

If it is clearly specified that the alternative method shall apply, a subcontract in the adapted form (NSC/4a) is to be entered into within 14 days of nomination, the full (basic) procedure above not being followed. In this shorter method an appendix in the contract incorporates essential details at the tendering stage (eg, retention, insurance, programme, daywork, fluctuations, set-off etc). NSC/2a (Employer/ Nominated Subcontractor Agreement) under this alternative method may be dispensed with, the proposed subcontractor to be informed accordingly at tendering stage (see 6.2.3 below for further details).

6.1.2 NON-NOMINATED ('DOMESTIC') SUBCONTRACTORS

The arrangements for subletting under clause 19 of the Standard Form (using the 'blue' form) are continued. A new class of selected subcontractor has been added, all subcontractors under this clause now being described as 'domestic':

(i) Where the architect requires priced work to be carried out by named firms at least three subcontractors must be listed by the architect in the bills from whom the contractor makes his selection, the work being fully measured and priced in the contractor's bills. If the number falls below three, it may be restored to this number by names being added by the employer or contractor with the agreement of the other. Additional names may be added by mutual consent. Failing that, if less than three names are available the contractor can proceed to sublet in the normal way under the terms of clause 19.
(ii) Existing arrangements continue for subletting to a subcontractor chosen by the contractor and approved by the architect (see also section 3.6.7).

6.1.3 PROVISIONS REVIEWED

It is proposed to deal with the new provisions in the following sequence:

Section 6.2: Consideration of the new and lengthy procedures for tendering and nomination which are designed to tighten up the business arrangements for letting nominated subcontract work and to define more clearly, the contractual relationships.

Section 6.3: The alterations necessary in the (main) Standard Form of Contract are examined with particular reference to clause 35 which now deals with the main contract aspect of nomination. Amendments to other clauses in the Standard Form were also necessary, but since these are not so directly related to the detailed arrangements for nomination as clause 35, they are incorporated in Chapter 3.
Section 6.4: The provisions of the new JCT Nominated Subcontract Form (NSC/4) are considered in detail. Since some of the clauses are similar to those in the 'green' form, a comparative schedule has been prepared to facilitate reference to the appropriate explanations in Chapter 5, thus avoiding repetition. Elsewhere in this book contract provisions have been considered under subject headings, grouped in the building sequence. NSC/4 is considered clause by clause because several clauses are similar to those dealt with in detail in other chapters.

It is hoped that analysing the provisions in this way will have the added merit not only of pin-pointing differences and enabling the reader to concentrate on those, but also of avoiding the need to study in detail clauses with which the reader may already be acquainted.

6.2 Tendering and nomination procedures

As already indicated there are now two alternative procedures for nominating a subcontractor and in respect of each nomination the basic method will apply unless otherwise stipulated at the tendering stage, or in any subsequent instruction requiring a variation or on the expenditure of a provisional sum leading to a nomination. Any subsequent change in method will be deemed a variation and valued accordingly but any instruction (to be valued) must be issued before nomination (see section 6.3.2). The basic method uses the Standard Form of Nominated Subcontract Tender and Agreement (Tender NSC/1) and the Employer/Nominated Subcontractor Agreement (Agreement NSC/2) followed by Nomination NSC/3 and the execution of the Standard Form of Nominated Subcontract (NSC/4).

The use of the alternative method must be stated explicitly in the tendering documents and also whether or not the employer intends to use Agreement NSC/2a. Within 14 days of the nomination instruction being given, the contracting parties must enter into the adapted form of subcontract (NSC/4a) (see section 6.2.3).

A point to note is that the Employer/Nominated Subcontractor Agreement is an essential part of the basic method, and an optional part of the alternative method.

The documents are examined in the next sub-section and this is followed by a summary of the procedures tracing the movement of the documents as the terms are agreed and finally signed.

6.2.1 BASIC METHOD DOCUMENTS

The basic method consists of the use of prescribed forms in a sequence set out in procedures recommended by the JCT. Only when the subcontractor has completed and executed NSC/1 and 2 may he be nominated. The forms are now considered with a description of the part played by the parties in securing the necessary agreement on a wide range of matters, before subcontract relationships are fully and finally established. An outline of the procedures then follows.

Tender (NSC/1)
NSC/1, which is to be used where a subcontractor is to be nominated, consists of a tender with two schedules which give full details of the basis of the offer. These agreed terms with annexed documents (drawings, bills, etc) together with NSC/4 will constitute the subcontract documents.

(i) *Tender*
This is the formal offer in response to an enquiry, after any pre-selection procedures have been completed, addressed to the employer and the main contractor. In it the subcontractor offers as a nominated subcontractor to carry out and complete defined work as part of the main contract in accordance with annexed signed and numbered drawings, specification, bills or rates as appropriate, for a VAT exclusive subcontract sum, or for a VAT exclusive tender sum where the work will be completely remeasured and valued (allowing for $2\frac{1}{2}$ per cent cash discount), and in accordance with the agreed particular conditions in Schedule 2 and the conditions in NSC/4. The daywork percentages are also set out in the tender.

The offer is subject to nomination within a period stipulated by the subcontractor from the date of his signature on the tender or any later date he may notify to the architect in writing. There is a right to withdraw within 14 days of written notification of the main contractor's identity (if not already known), and also if there is failure to agree with the main contractor the particular conditions in Schedule 2 (see below). In these circumstances the employer would only be liable to pay for any design work or materials already ordered under the terms of NSC/2 (see below). The tender is signed by the subcontractor, the architect on behalf of the employer (signifying approval) and the main contractor (as accepted subject to a nomination instruction), but it is made clear in a stipulation to the tender that it is not binding on the subcontractor until the employer has signed NSC/2 and the tender has also been signed as approved on his behalf.

(ii) *Schedule 1 – particulars of main contract and subcontract*
This gives basic information on the contract-employer, architect, quantity surveyor and main contractor (if then appointed). There then follows detailed information on subcontract conditions, Subcontract NSC/4 to operate unamended and to be executed under hand or under seal as appropriate, once nomination on NSC/3 has taken place. Fluctuations under clause 34 of NSC/4 may be on the basis of clauses 35 or 36 and there is an appropriate appendix for completion in respect of materials fluctuations including electricity, and fuels where appropriate. Where the formula is used under clause 37 the appendix gives full details of the general or specialist formula rules to apply, the non-adjustable element, base month, etc. Clause 35 must

be used if neither clauses 36 nor 37 apply.

Since the subcontract (NSC/4 recital 4) declares that the subcontractor has had notice of all main contract conditions set out in the schedule, this must be comprehensive and accurate; it takes precedence where there is conflict between other subcontract provisions and the main contract. Since the schedule therefore sets out main contract provisions, it gives details of main contract works, the form of main contract used (eg, 1980 edition, private or local authorities, with or without quantities) whether to be executed under hand or under seal and whether amended in any way. Main contract alternative provisions must be identified: eg, fire insurance – clause 22; and any provisional sum under clause 21.2.1 (joint named cover public liability insurance).

The main contract appendix, duly completed, is given in full and the terms where relevant will apply to the subcontract unless otherwise stated (eg, retention and defects liability period).

The place must be given at which unpriced bills, drawings, etc may be inspected. Any requirements of the employer on the order of the works, location and type of access, any obligations or restrictions imposed by the employer and not covered by the main contract conditions, together with any other relevant information must also be set out. This comprehensive statement, which is to be prepared by the architect and later checked by the contractor, must be signed by the subcontractor as 'information noted'.

(iii) *Schedule 2 – particular conditions*

Once the contractor receives NSC/1 and the architect's preliminary notice of nomination he must agree the terms of Schedule 2 with the proposed subcontractor. As a preliminary indication to the architect and the contractor, the subcontractor sets out details of his preliminary programme requirements subject to any stipulation on timetable by the architect, including the period (excluding time for approval) for preparatory work on further design, working and shop drawings, the period (off-site and on-site) for executing the works and the notice required to commence. Programme details (including completion date) and the order of works once agreed must be clearly set out and initialled by main contractor and subcontractor, the subcontractor's preliminary proposals (1A, 1B, 5A and 9A) then to be deleted.

The provision for attendance to be provided free of charge to the nominated subcontractor calls for special comment. General attendance must be given as a separate item in each case in the bills and priced accordingly (SMM B.9.2). Other (special) attendance varies immensely depending on the nature of the specialist work involved, and items must be detailed in the bills for separate pricing by the contractor in accordance with SMM 6th edition B.9.3. (See also NSC/4 clause 27 references in section 6.4.6.) Schedule 2 of NSC/1 deals with this in two stages. The subcontractor in his tender sets out in section 3A his attendance proposals under the seven SMM headings (special scaffolding; access roads and hardstanding; unloading; hoisting and placing; covered storage including light and power; power supplies; temperature or humidity levels, and any other requirements not otherwise covered). These are then agreed between the main contractor and the subcontractor any agreed alteration to be set out in Section 3B.

Other matters detailed are agreed insurance requirements with any indemnity limits, industrial agreements (to be finalised), adjudicator and trustee stakeholder (for set-off provisions see clause 24 NSC/4), information under the statutory tax deduction scheme (status of main contractor and sub-contractor, and on subcontractor's certificate the date of expiry and evidence required) whether VAT clause 19A or 19B applies and whether NSC/4 is to be executed under hand or under seal. Any other matters such as limitation on working hours must also be given. The subcontractor undertakes in NSC/2, once preliminary notice of nomination has been given, to seek to reach agreement on all these points with the main contractor, details to be recorded and signed by both parties. The architect is then to be informed through the main contractor.

If the particular conditions in Schedule 2 cannot be agreed within ten days (eg, on other attendance proposals, or on the starting date or period to carry out the work) the architect must be informed at once so that the difficulty can be resolved. Failure to agree could result in the subcontractor withdrawing his tender.

Tender NSC/1 is published in pads of three in a set, with notes on the front entitled 'Notes on the use of Tender NSC/1 and Agreement NSC/2', (see section 6.2.2). Agreement NSC/2 is published separately .

Employer/Nominated Subcontractor Agreement (NSC/2)

There are advantages in a separate agreement between employer and nominated subcontractor which creates a contractual relationship between them. This is particularly so where a design element is involved but in addition direct responsibility can be placed on the subcontractor in specific circumstances where he is the cause of delay to the main contract works. The subcontractor can also benefit from undertakings on payment. This was recognised in the optional RIBA Employer/Subcontractor Agreement, the 1973 edition being reviewed in section 5.12.

In the new JCT Nominated Subcontract Form this type of agreement (NSC/2) is a separate but integral part of the basic method documentation and is used in conjunction with the tender. Although the architect is named, he is not a party to the agreement (or the tender) nor in any way liable to the subcontractor on the matters referred to. Similarly the main contractor is not a party to NSC/2 although, for information purposes only, he is given a copy of the agreement.

The following are the terms agreed between the parties, after the tender has been submitted and approved by the architect who, once there is agreement on all matters (including NSC/1 Schedule 2), will issue a nomination instruction on NSC/3.

Obligation of subcontractor

Once the architect has issued his preliminary notice of nomination and the 'particular conditions' in Schedule 2 of NSC/1 have been agreed and signed by the main contractor and subcontractor (the architect being informed), the subcontractor then warrants that reasonable skill and care have been and will be exercised, as appropriate, on design, selection by him of materials and goods and in any prescribed performance specification (see section 6.3.10).

After the date of this agreement, but before nomination instruction NSC/3 is issued, the subcontractor may unless he so qualifies his offer, be instructed in writing by the architect to proceed under the agreement with any design work or the ordering of materials for the works, the employer to pay for this in advance of a nomination. Materials paid for become the property of the employer; design work is to be used only for the purpose of the works. After nomination on NSC/3, payments are to be in accordance with the subcontract, any earlier contractual payment to be credited. However it there are materials or design work previously ordered but subsequently not used for the works by a written decision of the architect these would be paid for separately.

Once nomination instruction NSC/3 has been issued the subcontractor undertakes to supply information (and drawings) in accordance with the programme subject to any extension of time under NSC/4, clause 11.2 or as and when reasonably required by the architect so that the architect's instructions or drawings under the main contract are not delayed thus giving the main contractor grounds for an extension of time under clause 25.4.6 or a loss and/or expense claim under clause 26.2.1.

Similarly, the works must be carried out so as not to cause the architect, by reason of the subcontractor's default, to issue instructions on:

(a) determination (and subsequent re-nomination) under clause 35.24 (suspension of work under clause 21.8 of NSC/4 excluded)
(b) an extension of time for the main contractor under clause 25.4.7 (delay on the part of a nominated subcontractor).

The subcontractor undertakes to indemnify the employer for any direct loss and/or expense suffered as a result of a re-nomination under clause 35.24 (except where the nominated subcontractor determines).

Where in a specified supply or subcontract (NSC/4 clause 2.3 – see section 6.4.5) there is any restriction, limitation or exclusion of liability, the subcontractor must inform the main contractor at once. He, in turn, will inform the architect and if both approve in writing the liability of the subcontractor to the main contractor will be limited accordingly. Under clause 35.22 of the main contract the liability of the contractor to the employer will be similarly limited.

Obligations to subcontractor
The employer specifically undertakes through his architect, to observe certain provisions of the main contract in regard to payment, failure to do so giving the subcontractor a direct right of action against the employer:

(a) Immediate notification in writing of value of subcontract work in interim certificates (clause 35.13.1)
(b) Early final payment under the terms of clause 35.17. (*Note*: Where he is responsible, under NSC/4 the subcontractor is to rectify at his own cost faults, defects, etc – up to the issue of the final certificate – or to meet the cost of doing so. Under NSC/2 5.3 he accepts the main contract obligations in

respect of his subcontract works for defects etc arising after its issue (see clause 35.18 (section 6.3.7) and NSC/2: 5.2 and 5.3).

(c) Payment direct by employer where contractor fails to pay (clause 35.13). The employer must be indemnified if in good faith he pays direct after bankruptcy or liquidation (see Standard Form clause 35.13.5.4).

It is made clear that this agreement prevails where its terms are in conflict with those of the tender.

Arbitration is provided for, with the stipulation that if the optional joinder provisions apply and the issue is substantially the same as one already referred to an arbitrator, the parties agree to refer the matter to this arbitrator (if suitably qualified) and be bound by his decision (see section 3.10).

Nomination – NSC/3

This is the briefest document in the series of four. It is the Standard Form of nomination signed by the architect under clause 35.10.2 where tender NSC/1 has been used.

It is addressed to the main contractor and identifies the main and subcontract works and the references in the bills of quantities or specification. The nomination refers to the date of the preliminary notice and the fully completed NSC/1 and sets out the name and address of the subcontractor formally nominated.

Subcontract conditions – NSC/4

In view of the extent and importance of the terms of the Standard Form of subcontract for subcontractors (NSC/4) a special section (6.4) is devoted to this. Before considering the clauses in detail, the terms of clause 35 of the Standard Form 1980 edition are reviewed in section 6.3 since these are closely interrelated.

6.2.2 BASIC METHOD PROCEDURES

To establish the terms of the subcontract there is a sequence of events designed to ensure that every point is agreed by the time the contract is made. The forms have already been described with some indication of the procedures involved. Below is a chronological summary tracing the interrelated actions leading to the making of the contract. This should be considered in conjunction with the chart on pages 128–30 and also the notes on the pads of NSC/1 Forms.

The subcontractor ('proposed' until binding arrangements have been completed) is first selected for nomination by the method chosen (eg, competition, negotiation, etc). The initial step must then be taken by the architect:

(1) NSC/1 is completed as far as possible by the architect, on behalf of the employer (one plus two copies) and sent to the proposed subcontractor along with NSC/2.

(2) The subcontractor completes his part of NSC/1, signs on page 1 (also in Schedule 1) and returns it (one plus two copies) to the architect along with NSC/2 duly executed, under hand or under seal as required.

(3) The architect has the tender (one plus two copies) signed on page 1 as approved by or on behalf of the employer and arranges for NSC/2 to be executed by the employer who keeps the original and gives a certified true copy to the architect which he sends to the subcontractor.
(4) The architect then issues a preliminary notice of nomination to the contractor together with NSC/1 (one plus two copies) as so far completed and a copy of NSC/2 as executed, for the contractor's information and retention.
(5) The contractor checks that NSC/1 is correctly completed so far, and then proceeds to settle the remaining terms with the subcontractor (ie, schedule 2 – programme, other attendance, etc). At that stage these are only proposals provisionally approved by the architect, and need to be discussed and finally agreed between the contractor and subcontractor. (If there are any problems a further instruction should be sought from the architect – see main contract clause 35.8 and .9.)
(6) Once Schedule 2 is completed and signed by both parties, the subcontractor must inform the architect through the main contractor. The original NSC/1 plus two copies is then returned fully executed to the architect by the contractor.
(7) All matters having been agreed and the necessary signatures obtained, the architect then issues his nomination on NSC/3 to the contractor with the original completed NSC/1, and a copy of NSC/3 to the proposed subcontractor with a certified copy of the completed NSC/1.
(8) The contractor and subcontractor execute NSC/4 (in duplicate) as provided for in the tender (ie, either under hand or under seal) and the contract arrangements are now complete, the parties to NSC/4 being only the main contractor and the nominated subcontractor.
(9) The documents are finally in the following hands:

	Architect	*Contractor*	*Subcontractor*	*Employer*
NSC/1*	Certified copy (or with the employer)	Original	Certified copy	
NSC/2	(Copy)	Information copy only	Certified copy	Original
NSC/3	(Copy)	Original	Certified copy	(Copy)
NSC/4	(Copy)	Original (executed in duplicate, each party retaining one)	Original	(Copy)

*with annexed documents
Note: (Copy) indicates likely business arrangement; it is not part of the official procedure.

6.2.3 ALTERNATIVE METHOD: DOCUMENTS AND PROCEDURES

In clause 35 there are separate provisions for the basic method (subclauses 5 to 10) and for the alternative method (subclauses 11 and 12). To recapitulate, the basic method is used unless the contract bills or an instruction for a variation under clause

13.2 or the expenditure of a provisional sum under clause 13.3 state that clauses 35.11 and 35.12 (the alternative method) shall apply. Any change in method must be before nomination, the instruction to be treated as a variation and valued accordingly (for full details see section 6.3.2).

There are only two documents in the alternative method – the Employer/Nominated Subcontractor Agreement and the Subcontract Conditions, given the references NSC/2a and NSC/4a. Neither Tender NSC/1 nor Nomination NSC/3 are used. In subcontract works not critical to the main contract programme and where subcontractors are selected after the main contract is let, this short procedure may be preferred. The proposed subcontractor must be informed at the tendering stage if Agreement NSC/2a will not be used. In making this decision the employer should remember the loss of remedies against the subcontractor particularly on design etc, faults and delays. The subcontractor, too, loses remedies against the employer on payment provisions. Dispensing with NSC/1 (tender form with accompanying schedules) means two things.

Firstly, the terms of the subcontract must be clearly set out at the tender stage and incorporated in an appendix to NSC/4a, since NSC/1 with its extensive details will not be used for completion and agreement before nomination.

Secondly, the references in NSC/4 to the tender and schedules must be altered in NSC/4a to references to an appendix and to the subcontract documents as appropriate. The significant amendments are dealt with below.

Since NSC/1 is not used the architect if seeking prices in competition would probably invite tenders from subcontractors in the 'old' way, ie, by using an enquiry form and certainly by setting out the full details on which tenders are invited, to be incorporated subsequently in the appendix to NSC/4a. Once a tender is accepted and NSC/2a executed where used, the architect under clauses 35.11 and 12 of the Standard Form issues a nomination instruction to the contractor (with a copy to the subcontractor). Within 14 days of this instruction the contractor must proceed to enter into subcontract NSC/4a with the proposed subcontractor.

The main features of these two documents are as follows:

NSC/2a – Standard Form of Employer/Nominated Subcontractor Agreement

This is very similar to NSC/2 except that the nomination is under Standard Form clause 35.11 and 12 and the recitals do not refer to NSC/1 and NSC/3 since these are not used. Instead there is a reference to the main contract and the architect, the tender by the nominated subcontractor (on the basis that NSC/2a will be used) the approval of the tender by the architect on behalf of the employer under clause 35.11 and 12, and the issue or intention to issue a nomination instruction. Finally it is made clear that the architect is in no way liable to the subcontractor under the agreement.

Clause 1 of NSC/2 is not appropriate to NSC/2a since it refers to the basic method.

Nominations are referred to as 'instructions nominating the subcontractor' (NSC/3 not being used). The subcontract references are of course to NSC/4a and its Appendix.

Apart from this the purpose and provisions of the agreement remain unaltered.

NSC/4a – Standard Form of Subcontract
This is entitled 'for subcontractors nominated under the Standard Form of Building Contract – clauses 35.11 and 35.12'.

The main changes from NSC/4 (see section 6.4) are at the beginning and the end of the conditions. Recitals 1 to 6 in NSC/4 are deleted and three brief ones take their place:

(1) The subcontractor has submitted a tender for the subcontract works referred to in Part 1 of the appendix and described in numbered documents annexed to NSC/4a under a main contract (detailed in Part 2 of the appendix) with name of employer.
(2) The architect has selected and approved the subcontractor (and where applicable Agreement NSC/2a has been entered into between the employer and the subcontractor).
(3) The architect has instructed on a nomination on the date set out, under clause 35.11: the contractor to enter into NSC/4a with the subcontractor within 14 days under clause 35.12.

In the articles the subcontractor undertakes to carry out and complete the subcontract works in accordance with NSC/4a and the annexed documents, the contractor to pay the subcontract sum (or the 'finally ascertained' subcontract sum) as set out.

A detailed appendix is necessary incorporating tendering details and giving most of the information NSC/1 would have provided. This is as follows:

Part 1 — Particulars of subcontract works, with listed annexed numbered documents.
Part 2 — Description of main contract works, details of main contract conditions, where documents can be inspected and whether the main contract is under hand or under seal. The alternative provisions of the main contract must be identified and any amendments noted. The details in the Appendix to the main contract must be given.
Part 3 — Extent of public liability insurance cover under clause 7.1.
Part 4 — Periods under clause 11.1 for preparation of further design work and working and shop drawings and the period (off-site and on-site) for execution and notice to commence.
Part 5 — Retention percentage (same as main contract) (clause 21).
Part 6 — The names of adjudicator and trustee-stakeholder (clause 24).
Part 7 — Details of other attendance (clause 27.2).
Part 8 — Fluctuations clause applicable (ie, 35, 36 or 37).
Part 9 — Clause 36 basic materials list; date of tender; percentage 36.8.
Part 10 — Clause 35 (as Part 9).
Part 11 — Formula details (clause 37) (rules, non-adjustable element, etc).

Apart from this, the amendments to the clauses in NSC/4 are minor. References to Tender NSC/1, etc are altered to 'the Appendix' or the 'Subcontract documents'

CHART OF JCT SUBCONTRACT BASIC METHOD PROCEDURES

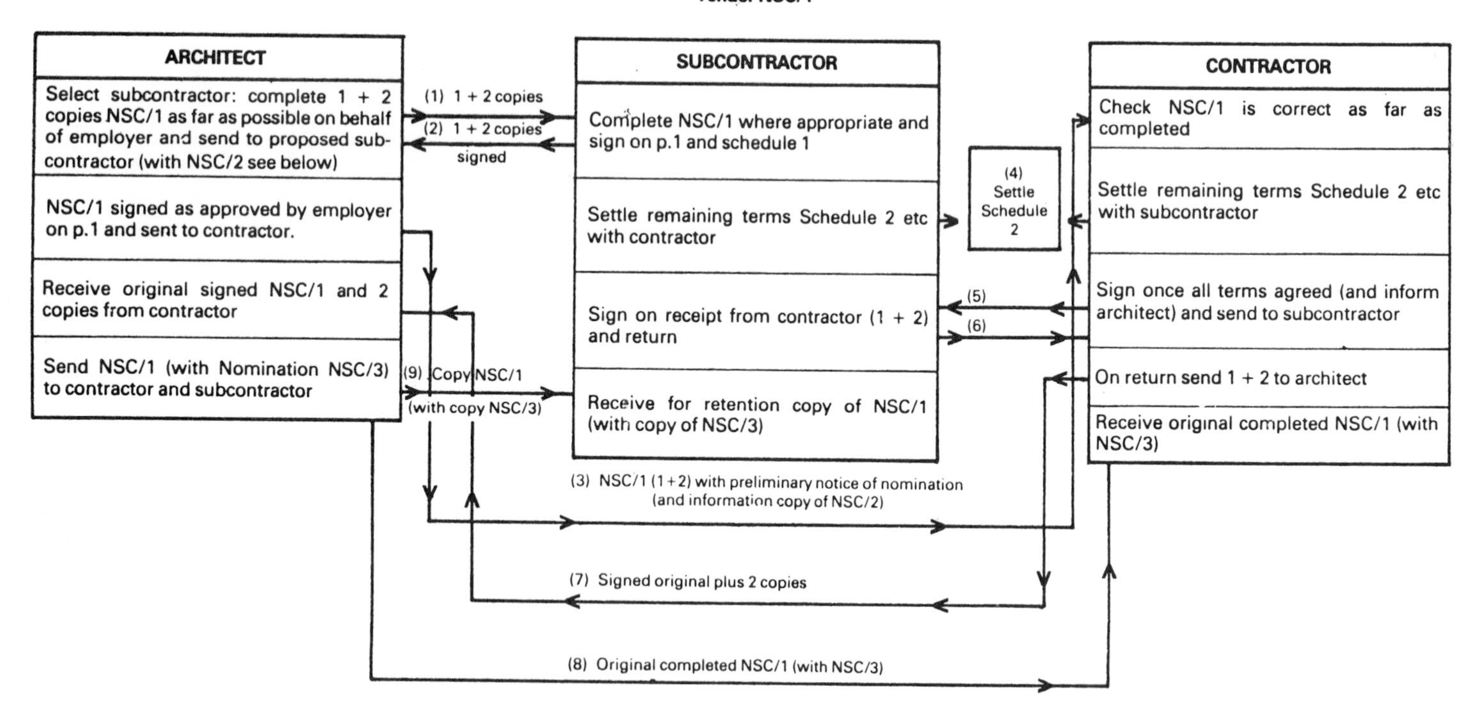

Agreement NSC/2

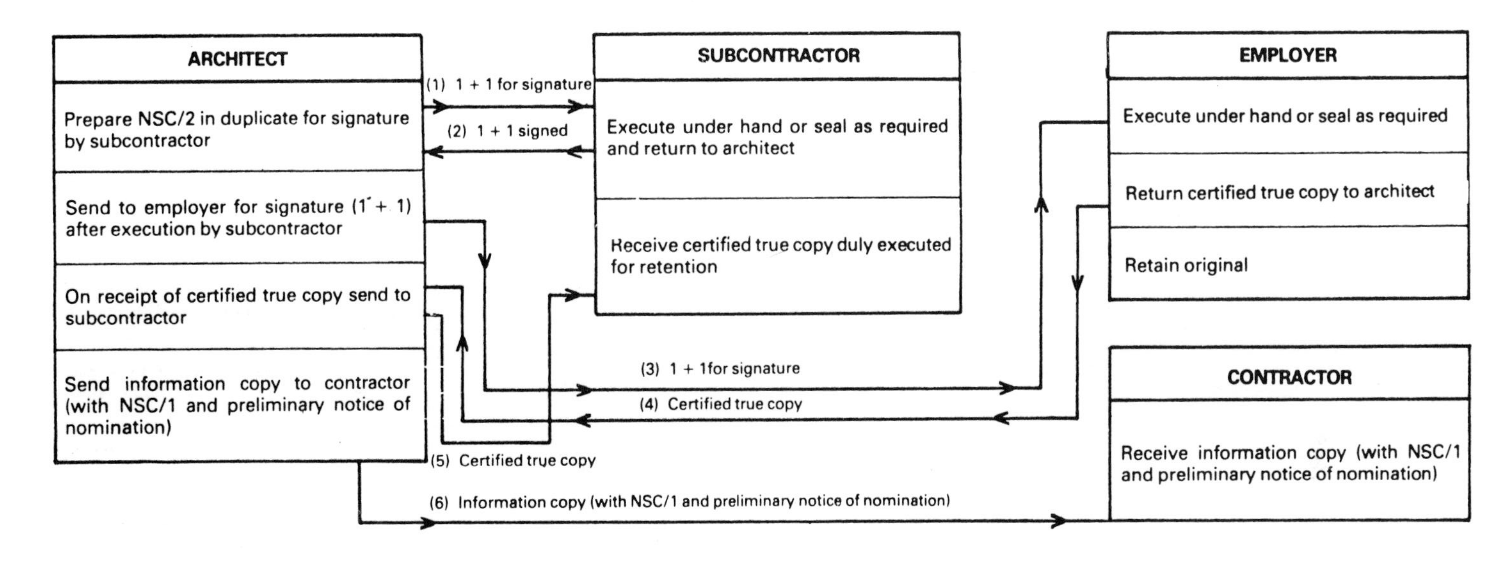

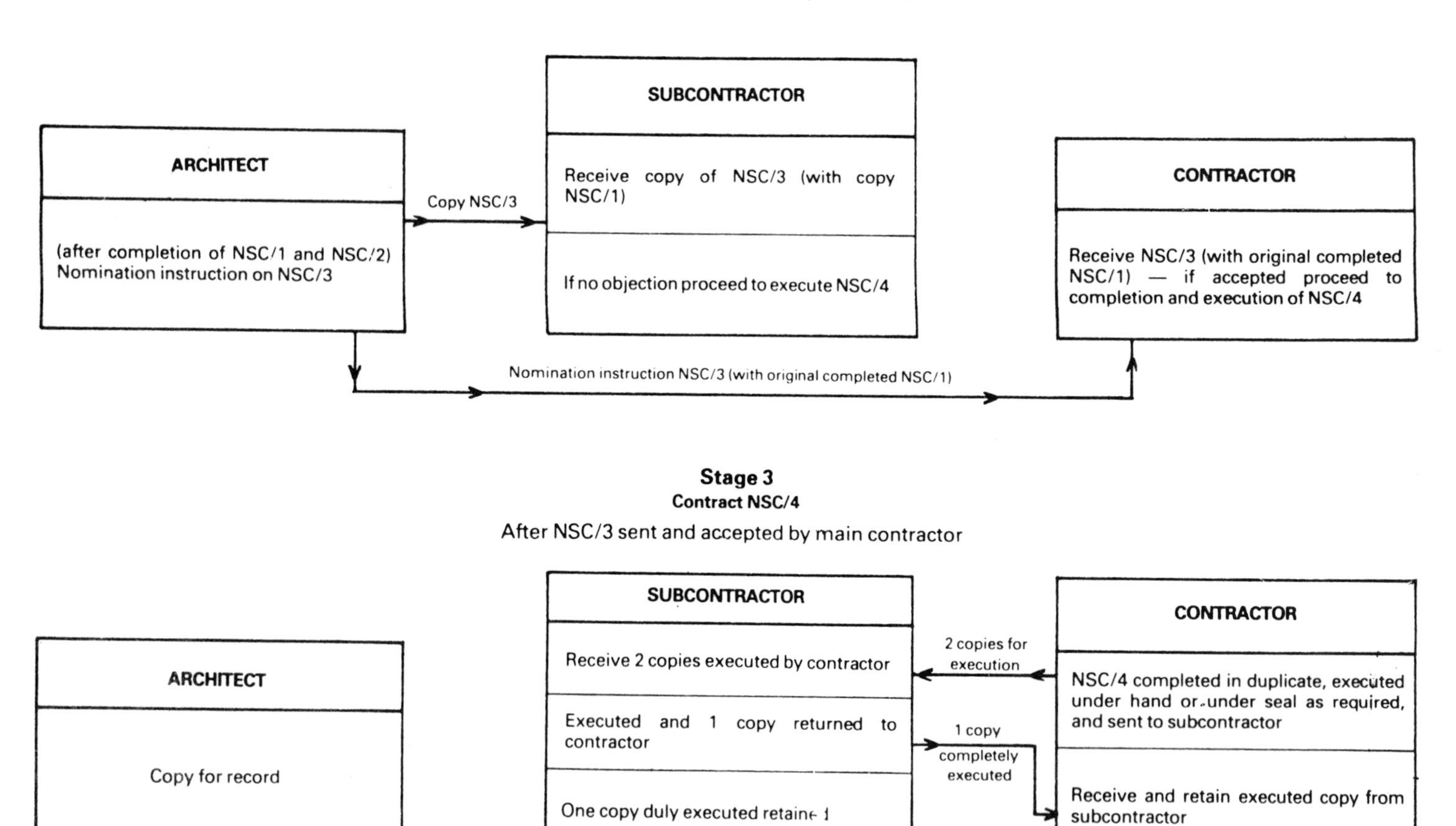
Stage 2
Nomination NSC/3
(after NSC/1 and NSC/2 fully executed)
ARCHITECT
(after completion of NSC/1 and NSC/2) Nomination instruction on NSC/3
Copy NSC/3
SUBCONTRACTOR
Receive copy of NSC/3 (with copy NSC/1)
If no objection proceed to execute NSC/4
CONTRACTOR
Receive NSC/3 (with original completed NSC/1) — if accepted proceed to completion and execution of NSC/4
Nomination instruction NSC/3 (with original completed NSC/1)
Stage 3
Contract NSC/4
After NSC/3 sent and accepted by main contractor
ARCHITECT
Copy for record
SUBCONTRACTOR
Receive 2 copies executed by contractor
Executed and 1 copy returned to contractor
One copy duly executed retained
2 copies for execution
1 copy completely executed
CONTRACTOR
NSC/4 completed in duplicate, executed under hand or under seal as required, and sent to subcontractor
Receive and retain executed copy from subcontractor

which comprise NSC/4a with numbered annexed documents. Daywork requires definition in the contract (it is detailed in NSC/1 where used) – clauses 16.3.4 and 17.4.3.

6.3 Provisions of clause 35 of the Standard Form of Building Contract – 1980 edition

As indicated at the beginning of this chapter the Standard Form of Building Contract has been substantially altered as a result of the introduction of the JCT Standard Form of Nominated Subcontract (NSC/4 and related documents) in January 1980. There are other amendments in the extensively revised 1980 edition of the Standard Form which directly relate to new provisions for nominated subcontractors, in particular clause 30 dealing wit . payment, retention, etc. Provisions on nominations of subcontractors and suppliers are contained in Part II of the Standard Form which is now divided as follows:

Part I – General-clauses 1 to 34
Part II – Nominated subcontractors and nominated suppliers – clauses 35 and 36
Part III – Fluctuations – clauses 37 to 40

The Standard Form 1980 edition is fully reviewed in Chapter 3. To present a complete picture of the new nominated subcontract arrangements clause 35 is specially considered here, the numbers in parenthesis being references to the subclauses which total 26.

Clause 35 starts off by defining a nominated subcontractor (subclause 1) and the documents relating to nomination (3). It then covers in detail the main contractor's tender for subcontract work (2), nomination procedures (4-12), payment (13), extension of time (14), failure to complete on time (15), practical completion (16), final payment (17-19), relationships with employer (20-22), nomination not proceeded with (23), re-nomination (24) and determination (25-26).

The documents and procedures referred to in subclause 4 to 12 have already been examined in detail. They are referred to briefly here since they are essential to a consideration of clause 35 of the main contract which ties in with the Nominated Subcontract Conditions (NSC/4 and 4a) dealt with in the following section. An adequate review of clause 35 must mean repeating some information on subcontracting provisions contained elsewhere in this chapter but clause 35 has been drafted to avoid repetition as far as possible and must be read in conjunction with the subcontract documents.

6.3.1 NOMINATED SUBCONTRACTOR – DEFINITION

Where an architect has reserved to himself the final selection and approval of a subcontractor to supply and fix materials or goods, or to execute work either by the use of a PC sum or by naming a particular subcontractor, by any of the means set out below, this subcontractor must be nominated in accordance with the provisions of clause 35 and will then become a nominated subcontractor for all the purposes of the contract. (See also the provision in SMM 6th edition B.9.1, clause 35.1 to apply notwithstanding ie, the right to nominate is not confined to a prime cost sum).

A nomination would follow from:

(i) a provision in the contract bills or
(ii) an instruction following the expenditure of a provisional sum in the bills under clause 13.3
(iii) a variation requiring work additional to that shown in the drawings or described in the bills, but of a similar kind to work for which a nomination was already provided in the contract bills
(iv) agreement (which cannot be unreasonably withheld) between the contractor and the architect for work so to be executed.

Work otherwise reserved for a nominated subcontractor may, if the architect is agreeable, be tendered for by the contractor (2). It must be work which is included in the contract bills and set out in the appendix, or arises from the expenditure of a provisional sum under clause 13.3, and of a kind which the contractor directly carries out in the ordinary course of business. If the tender is accepted the work must not be sublet without the consent of the architect. It will not be considered as nominated subcontract work under clause 35 and where this work arises from an instruction on the expenditure of a provisional sum, valuation will not be related to the valuation rules in clause 13 (variations) but to the accepted tender of the contractor (see clause 13.4.2).

6.3.2 DOCUMENTARY PROVISIONS

The documents related to nominated subcontractors are fully considered in section 6.2. along with the prescribed procedures.

However, to avoid cross-referencing and to assist in an understanding of clause 35, the documents listed in subclause 3 and the procedures in subclauses 5 to 12 are summarised below insofar as they relate to the provisions of clause 35.

There are two procedures: the basic method using the Tender NSC/1 and the Agreement NSC/2 leading to nomination on NSC/3 and the contract terms in NSC/4 (subclauses 5-10); and the alternative (and shorter) method using NSC/2a and NSC/4a (subclauses 11 and 12).

Basic method (5 to 10)

(i) NSC/1 and NSC/2 will apply to each nominated subcontractor unless the alternative method is clearly stipulated at the outset and if so whether or not NSC/2a will be used (ie, in bills, as a variation, or on the expenditure of a provisional sum).

(ii) A subsequent instruction changing the procedure either way will be considered a variation and valued accordingly under clause 13.2. However such an instruction can only be issued before preliminary notice of nomination (7.1 – basic method) or a nomination instruction

(11 – alternative method) except for work the subject of a renomination.

(iii) Before nomination, preliminary notice of nomination is given to the contractor along with the Tender NSC/1 so far as completed and a copy of NSC/2 duly executed, with instructions to settle with the subcontractor outstanding points in Schedule 2 of NSC/1.

(iv) The contractor then seeks the subcontractor's final agreement of all the particular conditions in Schedule 2 of NSC/1. If there is failure to agree within ten days, the contractor informs the architect and awaits further instructions.

(v) If the NSC/1 offer is withdrawn, the contractor must inform the architect at once and await further instructions.

(vi) Once all terms are agreed, NSC/1 duly completed (including Schedule 2) is sent to the architect who must at once nominate on NSC/3.

Alternative method (11/12)
Where it is expressly stated in the bills, or in an instruction leading to a nomination that the alternative method (ie, subclause 11 and 12) will apply and that NSC/2a will be used the architect can only issue a nomination instruction to the contractor (with a copy to the proposed subcontractor), once NSC/2a has been executed.

Within 14 days of the nomination the contractor must enter into the subcontract (NSC/4a) with the subcontractor.

Nomination procedure
An architect can only nominate a subcontractor who has tendered on NSC/1 and entered into agreement NSC/2, unless the alternative method under subclause 11 and 12 is indicated at the outset (6). No subcontractor can be nominated against whom the contractor has a reasonable objection, which must be made at the earliest practicable moment, but in any event not later than the date when the contractor sends the finally agreed tender (NSC/1) under subclause 10 to the architect. Under the alternative method where subclause 11 and 12 is used, the time limit under subclause 11 is not later than seven days after receiving the architect's nomination instruction (4). Where a reasonable objection is sustained, *or* where the subcontractor on the basic method does not agree with the contractor the particular conditions in Schedule 2 (NSC/1) within a reasonable time as requested, *or* where on the alternative method the nominated subcontractor fails without good cause to enter into subcontract NSC/4a (14 days is prescribed under subclause 12) another subcontractor must be selected for nomination by the architect or the nominated work omitted (23).

6.3.3 PAYMENT
Under clause 35.13 the architect 'directs' the contractor as to the amount of each

interim or final payment included in interim certificates under the main contract in respect of each subcontractor, computed and to be paid in accordance with NSC/4 or 4a as appropriate. Each subcontractor must be informed individually by the architect of amounts so included. (See section 6.4.6 for the payment provisions of NSC/4 clause 21 and section 3.8.2 for the terms of main contract clause 30 on payments.)

Before the issue of each interim certificate and the final certificate, the contractor must furnish reasonable proof to the architect that all sums included in previous interim certificates for the subcontractor have been discharged (13.3). If he fails to do so, the architect must issue a certificate to that effect with a copy to the nominated subcontractor. Where NSC/2 or 2a has been executed the employer must (and otherwise 'may') deduct any amounts for which the contractor has not furnished reasonable proof of discharge, from any payments due or to become due to the contractor including the final certificate and subject to any amounts due to the employer under the contract must pay these direct within 14 days to the subcontractor out of amounts available (13.5). Where there is no Agreement between the employer and the subcontractor this action is at the option of the employer.

If however the subcontractor fails to produce the evidence of payment reasonably required under the contract (eg, a statement or receipt) and the architect is reasonably satisfied that this is the sole reason for lack of proof, these provisions for payment direct do not apply (13.4).

The obligation to pay direct is limited to the amounts due on certificates and when that amount is in respect of retention, deductions will only be made up to the amount of the contractor's share. Where the employer has to pay two or more subcontractors, and the amount due or to become due to the contractor is insufficient to meet this in full, the payment to subcontractors is to be pro-rata to the amounts undischarged or some other basis the employer considers fair and reasonable (13.5.4.3). The provision for payment direct immediately ceases on the bankruptcy or compulsory liquidation of the main contractor. The employer must watch this point; any direct payment to a nominated subcontractor made in good faith after that date is recoverable under NSC/2 or 2a.

6.3.4 EXTENSION OF TIME

The granting of extensions of time by the contractor is provided for in NSC/4 (or 4a) but the written consent of the architect must first be obtained. The request to the architect for his consent must be accompanied by sufficient particulars of the expected effects and extent of the delay on the subcontract completion. Clause 11.2.2 of NSC/4 and 4a deals fully with the procedures including the provision that within 12 weeks from the request and supply of information a decision must be made on the extension of time (14) – see section 6.4.6.

6.3.5 FAILURE TO COMPLETE ON TIME

If a nominated subcontractor fails to complete the works, or any agreed part, within the period specified in the subcontract (extended as above) and the contractor notifies the architect (with a copy to the subcontractor), then provided all subclause 14 procedures have been dealt with the architect must, within two months of the date of the contractor's notification, so certify in writing to the contractor, with a duplicate to the subcontractor (15). (See also NSC/4 clause 12 – section 6.4.6.)

6.3.6 PRACTICAL COMPLETION

The architect must certify when the practical completion of the subcontract works is achieved, a duplicate copy of the certificate to be sent by him to the subcontractor (16). (See also NSC/4, clause 14 – section 6.4.6.)

6.3.7 FINAL PAYMENT

On meeting the conditions set out below, after the issue of the certificate of practical completion of the subcontract works the architect may, and after 12 months *must*, issue an interim certificate directing the contractor to pay the nominated subcontractor the final amount due to him under NSC/4 or 4a including the balance of retention. Before the final payment is certified the subcontractor must be informed of the computation of the final contract sum through the contractor.

This is subject to the following conditions:

(a) Agreement NSC/2 or 2a is in operation including clause 5.
(b) In the opinion of the architect and the contractor all defects under NSC/4 or 4a have been remedied.
(c) All necessary documents for final payment have been sent through the contractor to the architect (17).

This is a special provision in the absence of which final payment would be made in accordance with Standard Form clause 30.7 and NSC/4 clause 21.

Notwithstanding early final payment to the subcontractor, the main contractor remains responsible to the extent set out in the contract for damage to the works, etc for which payment has been made, up to the date of the practical completion of the main contract works or the date of possession by the employer and the fire, etc insurance provisions under clause 22 remain in full force (19). Defects, shrinkages or other faults as defined in NSC/4 or 4a, which appear in the nominated subcontract works after the final payment to the subcontractor has been made, are dealt with under subclause 18.

Thus if the nominated subcontractor fails to rectify defects appearing before the issue of the final certificate (clause 30.8), the architect must issue an instruction nominating another person (the 'substituted' subcontractor) to carry out the work of rectification to whom the clause 35 provisions will apply. The employer must take all reasonable steps under NSC/2 or 2a (clause 5.2) to recover the cost from the defaulting subcontractor but the contractor is responsible for the difference, if any, provided that before the further action was taken for remedying defects he agreed the price given by the 'substituted subcontractor'.

On the issue of the final certificate the architect's functions would be discharged and any defects appearing thereafter would be a matter between the employer and the main contractor. Under NSC/2 (and 2a) clause 5.3, the subcontractor is responsible to the employer and main contractor for the subcontract works to the same extent as the main contractor is responsible to the employer under the main contract.

It should be specially noted that the contractor's liability for defects does not extend to matters of design, materials, or performance specification dealt with under clause 2.1 of NSC/2 or clause 1.1 of NSC/2a to which only the employer and the nominated subcontractor are parties.

6.3.8 RE-NOMINATION

The circumstances in which a further nomination is necessary where the original nomination does not proceed are dealt with in clause 35.23 (see section 6.3.2 above).

Re-nomination is necessary in the following circumstances under clause 35.24, these provisions to be read in conjunction with clause 29 of NSC/4.

1 *Default under NSC/4 (or 4a) clause 29.1.1 to .4*

These matters are detailed in section 6.4.6.

The contractor must inform the architect when in his opinion there is default as specified on the part of the nominated subcontractor whose observations must be passed on to the architect. If the architect is reasonably of the opinion that there is default he must instruct the contractor to give notice to the subcontractor specifying the default. If under NSC/4 clause 29.1, the contractor wishes to determine, following notice of default, the architect must first instruct if he so requires. Thereafter the architect must make a further nomination for the completion of the subcontract works. If the default relates to defects, the contractor must agree the price to be charged for remedying these, as provided in clause 35.18 (see section 6.3.7 above).

2 *Bankruptcy or liquidation*

When the circumstances defined in clause 35.24.2 occur, the architect must make a further nomination. Where a receiver or manager is appointed the architect may postpone a nomination if there are reasonable grounds for supposing that the receiver or manager will continue with the subcontract in a way which will not prejudice the interests of any other party (ie employer, main contractor or any subcontractor).

Amounts payable to a subcontractor nominated under 1 and 2 above would be included in interim certificates and added to the contract sum (ie the contractor is not liable for additional costs resulting from renomination).

Note: At each stage the architect must be kept informed, his instruction being required for determination (25).

3 *Determination by nominated subcontractor under NSC/4 (or 4a) clause 30*

Since the main contractor would be at fault, any additional costs resulting from renomination would be met by the contractor, the employer deducting from monies due or recovering as a debt.

6.3.9 CONTRACTS OF SALE, ETC.

If the nominated subcontractor under NSC/4 or 4a clause 2.3, is required by the architect to enter into a contract of sale (or sub-subcontract) with any limitation of liability, this limitation will apply equally to the contractor in relation to the employer (22).

6.3.10 EMPLOYER/CONTRACTOR RELATIONSHIPS

The only liability of the employer to the nominated subcontractor is as set out in NSC/2 or 2a (20). The contractor is not responsible to the employer for any matter to which clause 2 of NSC/2 or 2a relates (eg, design etc) whether a nominated subcontractor is liable under this clause or not (21). The remedies of the employer against the subcontractor under clause 2 are so important as to make the optional use of NSC/2a most desirable.

6.4 Subcontract conditions – NSC/4

The JCT Nominated Subcontract Conditions (NSC/4) are part of a sequence of documents and the last in the process of clearly establishing main contractor/nominated subcontractor relationships before the contract is formally made.

The purpose of each document (NSC/1, 2, 3 and 4) and its place in the administrative procedures has already been explained in section 6.2 above but the lengthy and final form NSC/4 requires detailed consideration. In doing so an endeavour is made to indicate any conditions which may be similar to those already known in the industry in either the 'green' or 'blue' (nominated and non-nominated) forms. This will avoid unnecessary study and also keep this chapter within reasonable limits by cross-references to other chapters where the subject matter is already fully explained. Certain provisions mirror those in the Standard Form of Building Contract 1980 edition and these are cross-referenced to Chapter 3. The abbreviated NSC references are used for the subcontract forms.

A fair number of clauses have been substantially redrafted so that they bear little relationship to the 'green' form which will not be used in contracts under the Standard Form 1980 edition, NSC/4 being the appropriate subcontract form. In addition to this, certain clauses are entirely new. The amendments to NSC/4 to adapt it to the short procedure (NSC/4a) are considered in section 6.2.3 above.

6.4.1 ANALYSIS OF CLAUSES

To give a brief panoramic view of the clauses and their relationship to the old 'green' form the following table has been prepared. In several clauses the resemblance is remote although the subject matter may be the same (eg, payments). Where the terms are similar the clause titles are printed in italics.

Since the clauses are considered in categories the section references in 6.4 are given against the clause numbers for ease of reference, ie:

- 6.4.**2** – general review
- 6.4.**3** – clauses from 'green' form
- 6.4.**4** – clauses from 1980 Standard Form
- 6.4.**5** – clauses covering new subject matter
- 6.4.**6** – clauses covering matters in the 'green' form

Clause no	Discussed in subsection of 6.4	Title	Clause no ('green' form)
1	–	Interpretation, definitions, etc	–
2	5	Subcontract documents (new)	–
3	5	Subcontract sum – additions or deductions – computation of ascertained final subcontract sum (new)	–
4	6	Execution of the subcontract works – instructions of architect – directions of contractor	2, 7
5	3	*Subcontractor's liability under incorporated provisions of the main contract*	3
6	6	Injury to persons and property – indemnity to contractor	–
7	6	Insurance – subcontractor	4
8	6	Loss or damage by clause 22 perils to subcontract works and materials and goods properly on site	5
9	3	*Policies of insurance*	6
10	3	*Subcontractor's responsibility for his own plant*	19
11	6	Subcontractor's obligations – carrying out and completion of subcontract works – extension of subcontract time	8
12	6	Failure of subcontractor to complete on time	8
13	6	Matters affecting regular progress – direct loss and/or expense – contractor's and sub-contractor's rights	8
14	6	Practical completion of subcontract works – liability for defects	9
15	6	Price for subcontract works	–
16	6	Valuation of variations and provisional sum work	10
17	6	Valuation of all work comprising the subcontract works (new)	–
18	5	Bills of quantities – Standard Method of Measurement (new)	–
19A	3	*Value added tax*	10A
19B	3	*Value added tax – special arrangement – VAT (General) Regulations 1972, Regulations 8(3) and 21*	10B
20A	3	*Finance (No 2) Act 1975 – Tax Deduction Scheme*	10C
20B	3	*Finance (No 2) Act 1975 – Tax Deduction Scheme – subcontractor not user of a current tax certificate*	10D
21	6	Payment of subcontractor	11
22	3	*Benefits under main contract*	12
23	3	*Contractor's right to set-off*	13A
24	3	*Contractor's claims not agreed by the subcontractor — appointment of adjudicator*	13B
25	3	*Right of access of contractor and architect*	14
26	3	*Assignment – sub-letting*	15
27	6	General attendance – other attendance, etc	16, 17
28	3	*Contractor and subcontractor not to make wrongful use of or interfere with the property of the other*	18
29	6	Determination of employment of the subcontractor by the contractor	20
30	5	Determination of employment under the	

Clause no	Discussed in subsection of 6.4	Title	Clause no ('green' form)
		subcontract by the subcontractor (new)	–
31	6	Determination of the main contractor's employment under the main contract	21
32	6	Fair wages	22
33	5	Strikes – loss or expense (new)	–
34	4	Choice of fluctuation provisions	–
35	4	Contribution, levy and tax fluctuations	23B, C, D, E,
36	4	Labour and materials cost and tax fluctuations	23A, C, D, E,
37	4	Formula adjustment	23F
Articles of Agreement (3)	2	Arbitration	24

6.4.2 GENERAL REVIEW OF PROVISIONS

The following points call for broad general comment:

Format

On looking at NSC/4 one immediately notices that the format is entirely different from the old 'green' form but very much in line with the Standard Form 1980 edition. It is much longer and apparently more complex than its predecessor.

It is made clear on the title page that it is 'the JCT Standard Form of Subcontract for subcontractors who have tendered on Tender NSC/1 and executed Agreement NSC/2 and been nominated by Nomination NSC/3 under the Standard Form of Building Contract (clause 35.10.2)'. This is its place in the sequence and there are no short cuts if the full procedure is used.

There then follow on the first few pages the contents, then the seven recitals and three Articles of Agreement, the attestation and 37 clauses with well over 100 subclauses.

The Articles of Agreement considered below now include the arbitration provisions as in the Standard Form, and clause 1 is devoted to interpretation and definitions.

Articles of agreement

It will be apparent at once that the layout of the Articles is very similar to that of the Standard Form 1980 edition. The names and addresses of the contracting parties (contractor and subcontractor) are first set out before proceeding to the recitals.

Recitals: The recitals reflect the procedures under the basic method necessary to establish a full contractual relationship (ie, NSC/1 completed, NSC/2 executed and NSC/3 issued):

(a) The subcontractor has submitted a tender on Tender NSC/1 which has been duly completed and signed, with NSC/4 to apply unamended.

(b) The contractor has retained the original tender with a certified copy to both subcontractor and employer. A further certified copy is appended to the Articles of Agreement.

(c) The subcontractor has been fully informed through NSC/1, Schedule 1 of the provisions of the main contract except detailed prices.

(d) The architect has nominated the subcontractor in Nomination NSC/3.

Information is also given on the holding of a current tax certificate by subcontractor and contractor under the subcontractors' tax regulations, and whether the employer is a contractor under these regulations. If the position set out in NSC/1 Schedule 1 item 10 and Schedule 2 item 7 has subsequently changed by the date of execution NSC/1 must be corrected to accord with the recital.

Articles: The subcontractor undertakes to carry out and complete the subcontract works in accordance with the tender, the contractor agreeing to pay under the terms of the contract (the alternative bases – subcontract sum and ascertained final subcontract sum – are set out – see Tender NSC/1 p.1 and clause 15 which is reviewed below in 6.4.6).

Arbitration: The arbitration provisions follow in Article 3 and these are markedly similar to those in the 'green' form (clause 24) and reflect the principles in Article 5 of the 1980 Standard Form. In brief, disputes or differences arising during, after or on the abandonment of the works, between the contractor and subcontractor, or any matter under the subcontract may by notice in writing by either party be referred to an agreed arbitrator, or failing agreement to one appointed at the request of either party by the President of the RICS or else the RIBA. Both parties must agree in writing if there is to be arbitration before completion except on issues of payment, practical completion, where the set-off procedures under clause 24 are in operation extensions of time under clauses 11.2 and 3, or on whether a subcontractor's objection is reasonable under 4.3.

There is the optional joinder principle common to all JCT 1980 conditions (see section 3.10 for full details), Article 3.2 providing that if the dispute or difference is substantially the same as one under a related contract, the parties agree that the issue be referred to that arbitrator whose decision will be accepted as final and binding. There is the usual proviso that either party may withhold agreement if they reasonably consider that the arbitrator appointed is not suitably qualified to decide the point at issue. The option is as set out in NSC/1 Schedule 1.

The attestation clause provides for the alternatives of execution under hand or under seal as indicated in NSC/1 Schedule 2 item 11.

6.4.3 CLAUSES FROM 'GREEN' FORM

As will be noted from the above schedule there are certain clauses which are virtually identical to those in the pre-1980 nominated ('green') form subject to provision for the new procedures – NSC/1 etc – where appropriate. These are brought together below, and will not be reviewed in this chapter since they are fully dealt with elsewhere, the section references being given. The comparative clause numbers given are those in the 'green' form. One or two slight differences are referred to in the notes which follow:

NSC/4 clause no	Green form clause no		Discussed in this book in section
5	3	Subcontractors liability under incorporated provisions of the main contract (see note (a) below)	5.8.1
9	6	Policies of insurance	5.8.2
10	19	Subcontractor's responsibility for his own plant	5.8.2
19A	10A	Value added tax (see notes (b) and (c) below)	5.6
19B	10B	Value added tax – special arrangement (see notes (b) and (c) below)	5.6
20A	10C	Finance (No 2) Act 1975 – tax deduction Scheme (see note (d) below)	5.7
20B	10D	Finance (No 2) Act 1975 – tax deduction Scheme – subcontractor not user of a current tax certificate	5.7
22	12	Benefits under main contract	5.8.1
23	13A	Contractor's right to set-off (see note (e) below)	5.5.1
24	13B	Contractor's claims not agreed by subcontractor – appointment of adjudicator (see note (f) below)	5.5.2
25	14	Right of access of contractor and architect	5.9.2
26	15	Assignment – subletting	5.9.2
28	18	Contractor and subcontractor not to make wrongful use of or interfere with the property of the other	5.8.2

Notes

(a) A provision in clause 5.1.1 requires the subcontractor specifically to comply with clause 6, 7, 9, 16, 32, 33 and 34 of the main contract apart from the general provisions referred to in NSC/1 Schedule 1.

(b) A new provision in clause 19A.1.2 and clause 19B.1.2 stipulates that to the extent that the supply of goods and services to the contractor becomes exempt from VAT after the date of tender the subcontractor is to be protected against loss of input tax on goods and services for the subcontract works. This follows the provision in the Standard Form clause 15.3.

(c) Under clause 19A.4.4 and 19B.4.4 cash discount disallowed by the subcontractor and paid to him by the contractor is ignored in calculating tax under clauses 19A.4.2 and 19B.4.2 and 3. (Tender NSC/1 Schedule 2 states whether clause 19A or 19B applies.)

(d) Under clause 20A.2, not later than 21 days ('green form' 14 days) before the first payment is due, the subcontractor must lodge evidence of his current tax certificate with the contractor. The status of the subcontractor (see NSC/1 Schedule 2 item 7) determines whether clause 20A or 20B is deleted. Clause 20A would not apply where there were permitted Inland Revenue 'self vouchering' arrangements.

(e) In referring to the contractor's right of set-off against subcontractor's retention, it is made clear that this is notwithstanding the fiduciary obligation under clause 21.9.1. Under clause 23.2.3 the contractor's notice of set-off intention must be given to the subcontractor at least 20 days before the money is due (clause 13A(2)(c) 'green' form – 17 days).

(f) An adjudicator with an interest in any of the contracts of the parties may still act if all agree in writing. Clause 24.3.1 makes it clear that in considering the three courses of action open to him, an adjudicator need not only select one, but may also decide on any combination of them. Where a trustee-stakeholder is a deposit-taking bank, clause 24.5.2 provides that the usual interest shall be paid on sums deposited from which its charges and any tax on interest may be deducted.

6.4.4 CLAUSES FROM STANDARD FORM OF BUILDING CONTRACT 1980

In certain of the NSC/4 clauses there are extracts from clauses in the main contract (eg, insurance, definition of relevant events, loss and/or expense claims, etc). These are referred to in the review of NSC/4 clauses below (section 6.4.6).

In addition, however, the fluctuations provisions are identical in principle to those in clauses 37 to 40 of the main contract which are reproduced in clauses 34 to 37 of NSC/4 with small consequential alterations (eg, written notice in clauses 35.4.1 and 36.5.1 to be given to the contractor not the architect).

In NSC/1 Schedule 1, item 2, there should be a clear statement of the fluctuations provisions which will apply to the subcontract – that is, clauses 35, 36 or 37. The percentage under clauses 35.7 or 36.8 must be given. Appendix (A/B) which sets out data for materials fluctuations should have attached to it transport charges under clause 36.1.5.

Clause 34.2 states quite clearly that where either clauses 36 or 37 are not used, clause 35 must be used.

Work sub-sublet must, under clauses 35.3 or 36.4, be on the same fluctuations provisions as the subcontract.

The procedure is that only the selected alternative clauses are bound into the contract (NSC/4) to form an integral part of it (formerly all alternatives were printed, those not selected being deleted).

Only two provisions call for special comment and apart from the note below the detailed explanation of Standard Form clauses 37 to 40 in Chapter 4 applies equally to clauses 34 to 37.

(a) *Subcontractor in default over completion (clause 35.4.7 and 36.5.7)*

There will be no adjustments to the subcontract sum for fluctuations under clauses 35 or 36 where they occur after the date for completion set out in tender NSC/1 Schedule 2.1.C or any extended time granted by the contractor with the written consent of the architect, who has issued a certificate under clause 12 certifying failure to complete on time. This is subject to provisions for extensions of time in clauses 11.2 and 11.3 being unamended and the architect consenting or not consenting (ie, making a decision) on every request by the subcontractor for an extension of time under clause 11.2.

(*Note*: Clause 37.7 provides for the freezing of the indices on formula adjustment as in Standard Form, clause 40.7).

(b) *Applying formula adjustment (clause 37.2/3)*

This clause deals with the situation where the subcontract is subject to formula adjustment but the main contract is not. Valuations of the subcontract works will be needed for the purpose of calculating the formula adjustment, the subcontractor to be entitled to make representations through the contractor on valuations which the contractor must pass on to the architect. The non-adjustable element in the local authorities edition of the Standard Form under clause 40 would apply to clause 37, but where clause 40 is not used the non-adjustable element must be that inserted in the Tender Schedule 1 Appendix F.

The Nominated Subcontract Formula Rules (Series 2) were substantially revised in 1980 for use with NSC/4, clause 37 (see section 4.4.1 for details). The Rules to apply (Part I – general or Part III – specialist formulae) are set out in Tender NSC/1 Schedule 1 Appendix (F) with relevant details and are referred to in clause 37.1. (*Note*: The 1977 series Rules still apply to the 'green' and 'blue' subcontract forms.)

These matters having been dealt with, the way is now open to consider in detail the subcontract provisions which are entirely new or differ so substantially from the 'green' form as to require full consideration. Various parts of these NSC/4 clauses reproduce some of the 'green' form's provisions and this is indicated where appropriate.

6.4.5 CLAUSES COVERING NEW SUBJECT MATTER

Clause 2– subcontract documents

The elaborate documentation in subcontracting arrangements under the 1980 edition makes it necessary to define the subcontract documents, ie: the Tender (NSC/1) comprising p.1 and the completed Schedules 1 and 2 with all referenced and annexed documents (including NSC/4). (See also clause 18 where bills of quantities are a subcontract document.) Nothing in descriptive schedules may impose obligations additional to those in the contract documents. The terms of NSC/4 prevail over other subcontract documents, apart from the terms of the main contract set out in NSC/1 Schedule 1 which in turn prevail over the other subcontract documents. The accuracy of the main contract details in NSC/1 Schedule 1 is therefore vital.

Under subclause 3, where the subcontractor under the subcontract documents is required to enter into a subcontract or a contract of sale other than with the contractor and the liability of his subcontractor or supplier is restricted, limited or excluded, he is afforded protection by the following prescribed procedure. He must at once inform the contractor in writing of the restriction, who in turn must send a copy to the architect. (See also NSC/2 clause 8.) On receiving the written approval of the contractor and the architect (through the contractor), without which he may refuse to proceed, the liability of the subcontractor to the main contractor is similarly limited, restricted or excluded. (This is similar in intent to the provisions on *sale* conditions affecting the main contractor under clause 36.5 – see nominated suppliers in section 3.6.6. but see also clause 35.22 – in section 6.3.9).

Clause 3 – subcontract sum: additions, deductions
As in the Standard Form (clause 3) this clause makes it clear that any adjustments of, additions to, or subtractions from the subcontract sum must be 'taken into account' in the computation of the next interim certificate. This extends to sums to be included in the computation of the 'Ascertained Final Subcontract Sum' (see clause 15 below). Such a provision avoids repetition in subsequent clauses under which the subcontract sum may be adjusted (eg, variations, fluctuations).

Clause 18 — Bills of quantities
This clause clearly stipulates that where bills are a subcontract document then subject to clause 2.2 the quality and quantity of the work must be as set out in the bills which unless otherwise expressly stated are prepared in accordance with the principles of the 6th edition of the Standard Method of Measurement. Any departure from this or any error in description, quantity etc must be corrected and valued as a variation.

Clause 30 – Determination, etc by subcontractor
Although the right to determine his employment under the contract has been available to the subcontractor under the non-nominated ('blue') form, this is the first time a nominated subcontractor has been able to do so under the nominated form. The clause bears some resemblance to clause 22 of the 'blue' form and as far as rights and liabilities on determination are concerned is in virtually identical terms to Standard Form clause 28 where the main contractor determines (see section 3.6.14).

If there is no other adequate recompense under the subcontract, the subcontractor may determine his employment where the main contractor without reasonable cause, either wholly suspends or fails to proceed with the main contract works so that the progress of the subcontract works is seriously affected.

The point about 'no other adequate recompense' must be linked with the proviso that notice should not be given unreasonably or vexatiously.

The procedure is as follows:

(i) Notice from subcontractor to contractor and copy to architect, by registered post or recorded delivery, specifying default.
(ii) If for 14 days after notice the default continues or if later the default is repeated, determination of employment may at once be given by registered post or recorded delivery.

The rights and remedies, without prejudice to others accrued, or to the subcontractor's liability to the contractor under clause 6 for injury to persons or property, are briefly as follows:

(1) The subcontractor must promptly remove his temporary buildings, plant,

equipment and any materials not paid for by the contractor, taking every care to avoid damage, etc for which he is responsible under clause 6.

(2) After taking into account amounts already paid, the contractor must pay the subcontractor for work completed (valued under clause 21) executed but not completed (valued under clauses 16 or 17), for sums ascertained for direct loss and/or expense (under clause 13.1), materials under a binding order, reasonable cost of removing temporary buildings, plant, etc and for direct loss and/or expense caused to the subcontractor by the determination.

Clause 33 – Strikes: loss or expense

At a time of industrial conflict this new provision is apposite. In brief, if the main or subcontract works are affected by strikes, lock-outs, etc affecting any trades employed or engaged in the preparation, manufacture or transport of materials required for the works, neither party can claim on the other for loss or expense. 'All reasonably practicable steps' must be taken by the contractor to keep the site open and available to the subcontractor, and by the subcontractor to continue with the subcontract works. This provision does not affect any other rights under the contract if such action occurs.

6.4.6 CLAUSES COVERING MATTERS IN THE PRE 1980 SUBCONTRACT CONDITIONS

Insofar as subcontract conditions need to cover such points as performance, variations, insurance, payment and related issues, there are clauses on similar matters in the old nominated ('green') form and the new JCT Nominated Form. But there the similarity ends and it will be found that many provisions in the JCT form owe more to the Standard Form 1980 edition than to the 'green' form. This is due to the provisions in the main form relating to nominated subcontracts reflecting new concepts, procedures and documentation as well as some redrafting which improves the subcontract provisions.

Since the clauses in the new form are in a logical sequence, it is proposed to take those which call for special consideration in their numbered order. It will be noted that reference has already been made to clauses virtually identical to those in the 'green' form (section 6.4.3) and to those provisions which are entirely new (clauses 2, 3, 18, 30 and 33) (section 6.4.5). The clauses dealt with in this section are 4, 6, 7, 8, 11, 12, 13, 14, 15, 16, 17, 21, 27, 29, 31 and 32 (see table in section 6.4.1).

Clause 4 – Execution of the subcontract works

The subcontractor's obligation is to carry out and complete the works in compliance with the contract documents using materials and workmanship of the quality and standards specified and also to conform with all reasonable and relevant directions and requirements of the contractor. Where the architect's approval on materials and workmanship is required, this must be to his reasonable satisfaction. This is similar in principle to clause 2.1 of the Standard Form (and clause 2 of the 'green' form).

For the first time there is a requirement on the subcontractor to keep a person in charge continually on site while work is in progress. Any direction by the contractor, and any architect's instruction through him, is deemed to have been given to the subcontractor (cf, Standard Form clause 10).

Subclause 2 to 5 on instructions etc, follow the lines of Standard Form clause 4. Thus, architect's written instructions (including variations) through the main contractor, as well as his own written directions, unless challenged under subclause 6, must be complied with at once by the subcontractor subject to the proviso in clause 4.3 referred to below. On failure to begin to comply within seven days of the receipt of a written notice the contractor, with the architect's permission, may employ someone else to comply, the costs to be recovered from the subcontractor (clause 4.5).

The 'drill' on oral instructions and directions not confirmed in writing by the contractor within seven days is as in the Standard Form – the subcontractor confirms in writing to the contractor within seven days; if no dissent in writing in further seven days, then the instruction or direction is effective. There is also the provision about retrospective confirmation up to final payment, in writing by the contractor, where the subcontractor has complied with an oral instruction.

Subclause 6 reproduces clause 7.2 of the 'green' form by which the subcontractor, through the contractor, can ask under which clause of the main contract the instruction of the architect is authorised. If on receiving the architect's written answer from the contractor, the subcontractor complies he accepts the validity of the instruction; if he does not, then before compliance he may challenge this by giving notice of arbitration via the main contract, the contractor to be indemnified in return for his name being used (cf, Standard Form clause 4.2). An important new provision in clause 4.3 stipulates that a subcontractor need not comply with a variation defined in clause 1.3 to the extent that he has a reasonable objection to compliance (eg, change in employer's requirements – see also Standard Form clause 4). A dispute on this is immediately referable to arbitration.

Clause 6 – Injury to persons and property

This is a new indemnity clause similar in concept to clause 20 of the Standard Form. In it the subcontractor is liable for, and must indemnify the contractor against, any expense, liability, loss, claim or proceedings arising from:

(1) liability under statute or at common law for personal injury or death to any person (ie, not only employees) arising from the works unless caused by the act or neglect of the contractor, any other subcontractor, the employer, their employees or agents;
(2) liability for damage to property, real or personal, arising from the works where the subcontractor, his employees or agents are at fault (exceptions to this are loss by fire, etc under clause 8.1.1.1 (clause 22A cover) and damage at the sole risk of the employer under the main contract).

Clause 7 – Insurance

The requirement to insure is without prejudice to the liability to indemnify under clauses 5 and 6. It extends to:

(1) Cover for personal injury or death of any person, as defined under clause 6 above. (For employees, cover must comply with the Employers Liability (Compulsory Insurance) Act 1969.)
(2) Cover in respect of damage to property where the subcontractor is responsible including the special requirements of NSC/1 Schedule 2 item 4. Cover must extend at least to the public liability indemnity limits under the Standard Form, clause 21.1.1 set out in NSC/1 Schedule 1 item 10.

Clause 8 – Loss by clause 22 perils
This clause appears unduly lengthy but this is due in part to detailed extracts from the Standard Form. Space is saved by a reference to 'clause 22 perils', defined in clause 1 of the Standard Form.

Firstly Standard Form clause 22A is quoted and it is then stated that where this operates the subcontractor is not liable for these risks, nor is he required to insure against them. The responsibility is that of the contractor. Then clauses 22B and 22C risks are set out and this is followed by a subclause which says that for these risks neither the contractor or subcontractor is liable and neither is required to insure.

The clause then stipulates that where Standard Form clause 22A applies, the contractor must pay to the subcontractor the full value of the loss to the works valued as provisional sum work under clause 16 or 17, including materials properly delivered to site.

Where Standard Form 22B and C apply full details of the loss or damage must at once be reported in writing to the contractor. The subcontractor must then 'with due diligence' make good and proceed with the completion of the works. The costs of the subcontractor are dealt with as follows:

Clause 22B loss: restoration, replacement, removal etc treated as a variation under clause 13.2 of the Standard Form and valued accordingly.

Clause 22C loss: where the main contract is determined – share of the payments in accordance with clause 28 of the Standard Form (excluding provision for loss or damage on determination); where the main contract continues, due share of monies paid to contractor under clause 22C.2.3.

Clause 8.3 sets out provisions (similar to those in clause 5b of the 'green' form) on loss or damage other than by fire under clause 8.1 and 8.2 above to subcontractors' materials properly on site during the progress of the work.

Except for materials fully finally and properly incorporated in the main contract works the subcontractor will be responsible for loss or damage to his materials on site unless caused by the negligence etc, of the contractor, any other subcontractor, the employer, their employees or agents.

Loss or damage to materials incorporated before practical completion of the subcontract works will be the responsibility of the contractor unless the subcontractor is at fault.

On practical completion of the subcontract works the contractor will be responsible for loss or damage to the works handed over, except where damage is caused by the subcontractor.

This does not affect the subcontractor's liability for defects (see clauses 14.3, 14.4) including those after early final payment (see Standard Form clause 35.16 to 19).

The subcontractor must of course comply with the requirements of the fire etc insurance policies of the contractor or the employer (clause 8.4).

This division of liability on materials has given rise to frequent questions: Although the clause is detailed, the definition of 'fully, finally and properly incorporated' gives scope for argument especially in electrical and heating installations. It would appear that to meet the definition, complete installation, connecting up and testing are required. An understanding on site can avoid disputes. A footnote to the clause directs the subcontractor's attention to the risk of theft and vandalism against which he might insure.

Clause 10.2 makes it clear that it is the sole concern of the subcontractor as to whether and to what extent he insures for loss or damage caused to or by his plant, tools and equipment and also to other property and materials not properly on site for incorporation (eg, bought early and stored until needed). The risks would be his, the contractor to be indemnified for any expense unless he caused the damage.

As already indicated reasonable evidence of proper insurance under the contract may be required by either party of the other (clause 9). The contractor may insure where the subcontractor fails to do so, recovering the cost from sums payable to the subcontractor.

Clause 11 – Subcontractor's obligations: carrying out and completion of subcontract works – extensions of subcontract time

Once the subcontractor receives notice to commence and having regard to any delay in being able to start (under subclause 2 – see below) he must carry out and complete the subcontract works, as detailed in the Tender Schedule 2 item 1C and reasonably in accordance with the progress of the main contract works.

Subclause 2 deals with extensions of the contract time. The procedure is as follows:

Once it becomes reasonably apparent that commencement, progress or completion of the works, or any part, is delayed or likely to be delayed, immediate written notice must be given to the contractor of the material circumstances and the causes, identifying those due to relevant events (defined in clause 11.2.3 – below) or to delays caused by the contractor or any of his subcontractors. The contractor must then pass this on to the architect along with any written submission by the subcontractor. In his notice the subcontractor must if practicable give in respect of each matter the expected effects and his estimate of consequent delays in completion and must keep this information up-to-date.

If on receipt of this information, and a request for his decision, the architect is of the opinion that, for these reasons, the work is likely to be delayed beyond the subcontract period (as already extended) he must give his written consent to the contractor fixing a revised period for the work by granting an extension of time which the architect considers fair and reasonable after taking into account any omissions since the last application. (An omission could lead to a subsequent reduction in the period fixed.) This decision must be conveyed by the contractor within 12 weeks of the subcontractor's notice or within the period for completion if less, stating with the agreement of the architect the matters taken into account (see also clause 35.14 of the Standard Form: section 6.3.4).

Relevant events are defined in clause 11.2.3 and they reproduce the provisions of clause 25.4 of the Standard Form (see section 3.6.9) with one important addition – the valid exercise by the subcontractor of his right to suspend work under clause 21.8. The procedures are also similar except that it is the contractor who, with the consent of the architect, grants the extension the architect considers fair and reasonable and it is he who must, in agreement with the architect, state which relevant events have been taken into account and state also the allowance given for the default of the contractor or any other of his subcontractors.

Not later than 12 weeks after practical completion of the subcontract works the contractor, with the consent of the architect may on finally reviewing all the factors fix an earlier or later completion date or confirm the existing period for completion (see also clause 25 of the Standard Form).

There is, of course, the standard provision that the subcontractor must do his best to reduce delay and do all that may be reasonably required of him to maintain progress.

On failure of the architect to give written consent within the period allowed, the subcontractor may proceed with arbitration through the main contract with the usual provisos (eg, indemnity to contractor): clause 11.3.

Clause 12 – Failure of subcontractor to complete on time

Where the subcontractor fails to complete the works (or any portion) within the prescribed period as extended under the preceding clause 11, the contractor must notify the architect with a copy to the subcontractor. The architect if satisfied must then certify in writing under clause 35.15 of the main contract, within two months of the notification, that the subcontractor has failed to complete within the subcontract period. The subcontractor, who receives a copy of the architect's certificate, is then liable to the contractor for loss suffered by this failure.

Clause 13 – Matters affecting regular progress: direct loss and/or expense

This clause envisages regular progress being materially affected in several ways with resultant direct loss and/or expense:

13.1 – where the causes lie with the employer or the architect

13.2 – where the contractor or any of his subcontractors disturb the progress of the subcontract (cf, 'green' form 8(c)(ii))

13.3 – where the subcontractor disturbs the progress of the main contract works including any part subcontracted (cf, 'green' form 8(c)(iii)).

Subclause 13.1: For the reasons set out in clause 26.2 of the Standard Form (see section 3.6.13) the main contractor has a claim for loss and/or expense where the regular progress of the works is disturbed. A similar right for the same reasons is given to the subcontractor under clause 13.1, the procedure through the contractor who must act under clause 26.4 of the main contract, being as follows:

(a) A written application must be made to the contractor by the subcontractor as soon as it reasonably becomes apparent that regular progress has been or is likely to be materially affected and that direct loss and/or expense has been or is likely to be incurred which would not be reimbursed under any other provision of the subcontract. This must be accompanied by such supporting information, together with details of loss and/or expense as the contractor is requested by the architect to obtain in respect of the claim.

(b) The contractor must then require the architect to operate clause 26.4 of the main contract so that the direct loss and/or expenses may be ascertained, (amounts to be included in the next interim certificate).

Subclause 13.2: A claim lies against the main contractor for any direct loss and/or expense where regular progress of the subcontract works (including any part subcontracted) is materially affected because of any act, omission or default of the contractor or any of his subcontractors (or sub-subcontractors), their servants or agents. Within a reasonable time of this becoming apparent the subcontractor must give written notice to the contractor. The agreed amount is a debt recoverable from the contractor.

Subclause 13.3: The obverse of clause 13.2 is delay to the main contract works by the subcontractor (or his subcontractors), the terms and procedures being similar. The contractor could also include in his claim any liability to his other subcontractors for loss, etc suffered by them for which he would in turn be liable. The agreed amount is recoverable as a debt or from monies due to the subcontractor (eg, by set-off).

These provisions are without prejudice to other rights and remedies.

Clause 14 – Practical completion of the subcontract works: liability for defects

In view of the importance of identifying the date of practical completion from which the defects liability period (normally six months) runs, the subcontractor may give written notice to the contractor of what he considers this date to be. The contractor at once passes it on to the architect with his observations (of which the subcontractor is sent a copy). The date for the purpose of the contract will in any event be the one identified in the architect's certificate of practical completion of the subcontract works (or part thereof on early possession): clause 14.1 and 2. (See also clause 35.16 of the Standard Form – section 6.3.6).

Clause 14.3 deals with liability for defects which must be made good at the cost of the subcontractor in accordance with the instruction of the architect or the directions of the contractor. Clause 18 of the Standard Form sets out the liability of the main contractor for defects, the subcontractor to be similarly liable for faults, defects, etc in the subcontract works due to materials or workmanship not in accordance with the subcontract, or frost occurring before practical completion. (This provision should be studied in conjunction with Standard Form clause 35 subclauses 17 to 19 (defects after early final payment) – see preceding section 6.3.7.) Defects in design, etc under clause 2 of Agreement NSC/2 (or 2a) are a matter between the employer and the subcontractor.

There is the usual proviso that if the architect instructs that the cost of making good defects shall not fall entirely on the contractor, he in turn must grant a similar benefit to the subcontractor (clause 14.4).

A final subclause 14.5 requires the subcontractor on completion to clear up the works and leave them tidy to the satisfaction of the contractor.

Clause 15 – Price for subcontract works

This clause distinguishes between the two alternative bases on which a tender may be submitted on NSC/1: ie, a VAT exclusive subcontract sum or a VAT exclusive tender sum where the works are to be completely remeasured and valued. The JCT has made this distinction because many nominated subcontracts may be invited on an approximate tender sum with complete remeasurement of the works executed to be valued on the subcontractor's schedule of rates, or bill rates where bills are a contract document. This leads on to clause 16 for variations on the former basis, and clause 17 where a complete remeasurement including variations, takes place at the end of the subcontract to establish the ascertained final subcontract sum. It should be noted that the definition of variations in clause 1.3 on the former basis relates to alterations, etc in 'design quality or quantity', but on the later because of complete remeasurement to 'design or quality' only.

Clause 16 – Valuation of variations and provisional sum work

This clause provides the basis for valuing all variations defined in clause 1.3 and for all work executed by the subcontractor under an architect's instruction on the expenditure of a provisional sum in the subcontract documents where the tender is on the basis of clause 15.1 (a VAT exclusive subcontract sum).

The valuation is carried out by the quantity surveyor (or if none, the architect) agreed amounts to be included in the next interim certificate. Where measurement for this valuation is required, the contractor must give the subcontractor the opportunity to be present to take notes and measurements. These measurements must be on the same principles as the bills (if any), with allowances for lump sum or percentage and preliminaries adjustments.

Unless otherwise agreed by the contractor and the subcontractor with the approval of the employer, valuation shall be on the following bases:

If the subcontractor includes in his tender a schedule of prices for measured work and/or a schedule of daywork prices, valuations should be on this basis (clause 16.2). If not, the principles of the rules in the main contract for additions, substitution or omissions will apply (for Standard Form clause 13 (variations) see section 3.6.4). Bill or other rates apply if relevant (ie, work similar in character and conditions with no significant change in quantity), they are used if not, as a basis; failing that, a fair valuation. Where work cannot be properly measured and valued daywork rates, as defined, will apply where appropriate. The percentages on prime cost are to be set out in page 1 of NSC/1, the prime cost definitions to be identified, eg, RICS agreements with the National Federation of Building Trades Employers (see section 8.7), the Electrical Contractors' Association, the Electrical Contractors' Association of Scotland and the Heating and Ventilating Contractors Association (clause 16.3).

All vouchers etc are to be delivered at the end of the following week to the contractor for verification and onward transmission to the architect.

As in the Standard Form clauses 13.5.5 and 13.5.6 there is a provision in clauses 16.3.5 and 16.3.6 for valuation where the conditions under which other work is executed are substantially changed, and where in prescribed circumstances a valuation under the rules does not fully reimburse the contractor (see section 3.6.4 for details).

Any effect upon regular progress resulting in direct loss and/or expense caused under this clause is covered by clause 13.1.

Clause 17 – Valuation of all works comprising the subcontract works

Where the alternative basis of pricing is used (a VAT exclusive tender sum where the works are to be completely remeasured and valued – see clause 15.2 above) the contractor must give the subcontractor the opportunity of being present when the quantity surveyor is measuring for valuation. Where bills of quantities are a subcontract document the measurement must be on the principles used in their preparation. The rules for valuation of all work including variations are as set out in clause 16 above.

Clause 21 – Payment of subcontractor

This clause embodies the principles of clause 30 of the Standard Form and incorporates certain of its provisions by reference. In addition it ties in with the provisions of Standard Form clause 35 on nominated subcontractors, subclauses 13, 17, 18 and 19 containing provisions on payment, which in turn the employer under NSC/2 specifically undertakes to observe.

In considering this clause it is advisable to read in conjunction with it these two clauses in the main contract (see sections 3.8.2 and 6.3.3 and 6.3.7).

Clause 21 can be broken down broadly as follows:

subclause	*subject matter*
2	payment – application and representations
4	interim certificates – ascertainment
2	offsite materials
5, 6	retention – rules
9	retention – fiduciary interest
3	payment by the contractor
10	final payment – clause 15.1
11	final payment – clause 15.2
7	dispute as to certificate
8	right to suspend work

These aspects of payment are now considered in detail.

(a) Interim certificates: The subcontractor can make representations on what is included for him under NSC/4 clause 21.4, in interim certificates under the main

contract. Thus when ascertaining amounts due in interim certificates under Standard Form clause 30.2 which include payments in respect of nominated subcontract work the subcontractor may request the contractor to make application to the architect on any item to be included and to pass on his written representations (clause 21.2).

The items to be included are broadly in line with Standard Form clause 30.2 (see section 3.8.2) subject to any agreement on stage payments, with amounts due under clause 21.4 to be ascertained up to not more than seven days before the date of the certificate and to include the following:

Subject to retention (21.4.1)

(i) total value of work executed including work under clause 16.1 and formula adjustment under clause 37, if applying

(ii) total value of materials reasonably and properly delivered to site and adequately protected

(iii) off-site materials if architect agrees, under clause 30.3 of the Standard Form, the relevant conditions of which the subcontractor must observe

Not subject to retention (21.4.2)

(iv) Defects not chargeable under clause 14.4, and any payments under clauses 6 and 7 of the main contract.

(v) reimbursement for loss and/or expense (clause 13.1)

(vi) any fluctuations additions (clauses 35 or 36)

(vii) amounts necessary to allow for subcontractor's 2½ per cent cash discount on (iv), (v) and (vi) (ie, one thirty-ninth).

Less any fluctuations allowable (clauses 35 and 36) plus one thirty-ninth.

Note: It has been a point of argument in the past as to whether, for example, the subcontractor must allow cash discount on fluctuations, without reimbursement. Under this provision these amounts referred to are now paid in full.

From all these are deducted retention on items (i), (ii) and (iii) – see below – and amounts included in earlier interim certificates.

(b) Final amount due: Clauses 21.10 and 21.11 set out how the final contract sum should be ascertained on the alternative bases under clause 15:

15.1 – VAT exclusive subcontract sum (valuation of variations under clause 16)

15.2 – VAT exclusive tender sum (valuation of all work under clause 17)

All the documents necessary for the final adjustment of the subcontract sum (21.10) or the ascertainment of the final subcontract sum (21.11) must be sent either before, or within a reasonable time after the practical completion of the subcontract works to the contractor, or if he so instructs to the architect or quantity surveyor.

Clause 21.10: Once these documents are available the quantity surveyor must prepare a statement of the final valuation of all variations under clause 16 and the architect must send a copy to the contractor and the nominated subcontractor for the final adjustment of the subcontract sum. This must be done normally in the six month period from practical completion, as set out in the NSC/1 Schedule 1 item 10, or when necessary to arrange for early final payment to the subcontractor under Standard Form clause 35.17.

Deductions from subcontract sum
(1) all provisional sums
(2) omissions as valued
(3) fluctuations allowable plus one thirty-ninth
(4) any other amounts required by the subcontract to be deducted from the subcontract sum.

Additions to subcontract sum
(1) amounts payable under clauses 6, 7 (Standard Form) or 14.4
(2) variations other than omissions valued under clause 16.3
(3) value of work executed by the subcontractor under a provisional sum or described as provisional in the subcontract documents
(4) direct loss and/or expense under clause 13.1
(5) fluctuations under clauses 35, 36 or 37
(6) any other amount required to be added to the subcontract sum under the subcontract
(7) one thirty-ninth of 1, 4 and 5.

Clause 21.11: The ascertained final subcontract sum under clause 15.2 is the aggregate of:

(1) amounts payable under clauses 6, 7 (Standard Form) or 14.4
(2) the valuation under clause 17
(3) direct loss and/or expense under clause 13.1
(4) amounts payable (or allowable) under fluctuations (ie, clauses 35, 36 or 37)
(5) any other amount to be included in computing the ascertained final subcontract sum
(6) one thirty-ninth of 1, 3 and 4.

In both cases before the architect certifies the final payment for the subcontract works under clauses 35.17 or 30.7 of the main contract, a copy of the computation on the above basis must be supplied to the subcontractor.

(c) Payments by the contractor: All payments will be certified in *interim* certificates under the main contract, because under clause 30.7 all payments to nominated subcontractors are cleared before the issue of the final certificate in respect of the main contractor (see below).

Clause 21.3 gives the contractor 17 days from the date of issue of an interim certificate to notify and pay the subcontractor the amount included for him, subject to any set-off under clause 23 – less 2½ per cent cash discount on the amount paid on time.

There is the general principle that the contractor can claim against the subcontractor where deductions have been made by the employer in payments to the contractor, for reasons attributable to some act or default of the subcontractor his servants or agents. An employer's deduction under the main contract would be passed on to a nominated subcontractor by deduction from sums due (or recovered as a debt) but only to the extent attributable to that subcontractor's default (clause 21.3.1.2).

The nominated subcontractor is entitled to be immediately informed by the architect under Standard Form clause 35.13.1.2 of the amount of any interim or final payment included for him in any interim certificate. The subcontractor, on discharge of amounts due, must at once give the contractor written proof (eg, a receipt) to meet the architect's requirement under clause 35.13.5 of the main contract (see section 6.3.3).

As indicated above, final payment is made to the subcontractor at the very latest not less than 28 days before the issue of the final certificate (clause 30.7). Standard Form clause 35.17 to 19 deals in detail with provisions for earlier payment and liability for subsequent defects in the subcontract work (see section 6.3.7). A defects indemnity is given by the subcontractor to the contractor under clause 21.3.2.2.

The provisions are financially interesting since the contractor is given three extra days from the due date for honouring main contract certificates by the employer, to meet his commitments to his nominated subcontractors. Even if the employer does not honour the main contract certificate on time, the main contractor's obligation to pay the nominated subcontractor is unaffected.

(d) Retention: Clause 21 has been kept reasonably brief by referencing certain provisions to clause 30 of the Standard Form (see section 3.8.2 – rules for ascertainment and treatment of retention).

On retention the right of the employer to deduct and retain retention is, under clause 21.5, on the same basis as set out in clause 30.2.3.2 of the main contract, the retention percentages being detailed in clause 30.4 (eg, briefly, unless otherwise stipulated 5 per cent; half released on practical completion, the other half at end of defects liability period once defects made good).

Clause 21.6 states that retention is subject to the rules set out in clause 30.5 of the Standard Form. Under clause 30.5.2 at the date of each interim certificate, statements must be sent to the main contractor and nominated subcontractors detailing retention held in respect of each (with a copy to the employer).

Under clause 30.5.3 of the Standard Form any nominated subcontractor can, in the private edition, request the employer to place retention deducted in a separate identified bank account.

Clause 21.9 states that the contractor's interest in the subcontractor's retention as defined in the required statement under clause 30.5.2 of the Standard Form is fiduciary as trustee for the subcontractor with no obligation to invest, retention to be paid over as and when released. The contractor must immediately set aside in a separate bank account and become the trustee of a sum equivalent to this retention if he seeks to include in a mortgage or charge any nominated subcontract retention. Once this amount so set aside is paid, the final payment due to the subcontractor is to be reduced accordingly.

A contractor may withhold a subcontractor's retention, in exercising his right of set-off under clause 23. If this is contested by the subcontractor it would appear that because of the fiduciary interest he is required to place this money on the due date for payment in an identified trust account, the ultimate destination of these funds to await a decision under the procedure in clause 24 (clause 21.9.2).

However if an employer exercises his right of deduction from a certificate which includes nominated subcontract retention, this must be paid in full to the nominated subcontractor unless the deduction is attributable to his default. In these circumstances the shortfall to this extent can be passed on. If the contractor is in default the resultant deduction and shortfall must be borne by him since he would have no recourse against any subcontractor.

These provisions now ensure very considerable protection for the retention money due to nominated subcontractors.

(e) Dispute as to certificate: There is the standard provision (as in 'green' form 11(d)) that if the subcontractor is dissatisfied with the amount certified or with the architect's failure to certify, he may proceed to arbitration under the main contract using the name of the contractor with his agreement and appropriate indemnities (clause 21.7).

(f) Suspension – non-payment: The principle in 'green' form clause 11(e) is reproduced in clause 21.8 with a significant difference – there is now a monetary claim against the contractor for loss, damage or expense resulting from the suspension.

The grounds for suspension are failure by the contractor to pay the subcontractor as required under the contract (subject to any right of set-off) followed by the subsequent failure by the employer to operate the provisions for direct payment under Standard Form clause 35.13.5 or if operating the provisions, to pay the whole amount due to the subcontractor within 35 days of the date of issue of the interim certificate on which the main contractor has defaulted. The contractor and the employer must then be given 14 days' written notice of the intention to suspend the further execution of the work. Reasonable suspension may continue until payment is made, this period being added to the period for the completion of the subcontract works and not regarded as delay for which the subcontractor is liable.

Clause 27 – General and other attendance

The provisions of clauses 16 and 17 of the 'green' form are brought together in clause 27 with some interesting additions.

In requiring the contractor to provide general attendance free of charge to the subcontractor, clause 27.1 lists the items in SMM 6th edition B9.2. Under this clause the subcontractor must clear away all rubbish during the course of the subcontract works and keep the access clear (see clause 14.5 on liability to clear up at practical completion).

The items of other (special) attendance which are detailed in NSC/1 Schedule 2 item 3 must be provided by the contractor free of charge (for bill provisions see SMM

6th edition B9.3).

Subject to the provisions for general and other attendance the subcontractor must provide erect maintain and subsequently remove workshops, sheds, etc at his own expense. Subclause 3 requires the contractor to allocate places for these (subject to any reasonable objection by the subcontractor) and to give reasonable facilities for erection.

Subclause 4 has mutual provisions on scaffolding. While erected the scaffolding of both contractor and subcontractor may be used by each other for the purposes of the works, any warranty on fitness, condition or suitability being expressly excluded.

Clause 29 – Determination of employment of subcontractor by contractor
It is made clear in this clause that an instruction of the architect is a necessary step in determination. In this and several other respects it differs from clause 20 of the 'green' form.

Under subclause 1 the defaults giving grounds for determination are similar to those in the Standard Form (clause 27.1) where the contractor is in default:

- (i) complete suspension before completion without reasonable cause (eg, not suspension under clause 21.9); or
- (ii) failure to proceed according to programme (clause 11.1); or
- (iii) refusal after written notice from the contractor to remedy defects under the contract, remove defective work or improper materials materially affecting the works, or
- (iv) failure to comply with clause 26 on assignment and subletting or clause 32 (fair wages).

The procedure is then for the contractor to inform the architect, with any written observations of the subcontractor, of the defaults. If he so decides the architect then instructs the contractor under Standard Form 35.24.4.1 to issue a written notice by registered post or recorded delivery specifying the default, with a copy to the architect. If the default continues for fourteen days after the notice has been given, or the default is repeated at any time thereafter, determination of the employment of the subcontractor on architect's instruction (Standard Form 35.25) may follow within ten days by registered post or recorded delivery. There is the proviso that notice must not be given unreasonably or vexatiously and that determination is without prejudice to other rights and remedies.

Under clause 29.2 there is automatic determination on bankruptcy, liquidation, etc (see Standard Form clause 35.24 section 6.3.8).

Clause 29.3 sets out the rights and duties of the contractor and subcontractor on determination (so long as the subcontract has not been reinstated or continued):

1. A subcontractor nominated to complete (see clause 35.24 of the Standard Form: section 6.3.8) may enter on the works; use the former subcontractor's temporary buildings, plant and site materials; purchase necessary materials to complete.
2. Except in bankruptcy or liquidation, the subcontractor must if required by

the employer or architect within 14 days of determination (the contractor consenting) assign to the contractor without payment the benefit of agreements for materials or work. The terms must be agreeable to the supplier or sub-subcontractor to whom outstanding payments may be made by the contractor as directed by the architect.

(3) On a direction of the contractor or an instruction of the architect the subcontractor must remove his temporary buildings, plant and materials. On a failure to comply promptly, the contractor may remove and sell (without being responsible for loss or damage) the proceeds to be credited to the subcontractor less costs incurred.

Under clause 29.4 the subcontractor must pay the contractor the amount of any direct loss and/or damage caused by the determination. No further payment is to be made to the subcontractor until the subcontract works are completed, when any application for payment would be forwarded by the contractor to the architect. From the amount certified as due up to the date of determination and not yet paid, the employer would deduct his expenses and direct loss and/or damage suffered. The contractor in paying the balance over to the subcontractor, subject to 2½ per cent cash discount, would make a deduction for any direct loss and/or damage suffered (without prejudice to any other rights).

Determination by the subcontractor – a new provision under clause 30 – is dealt with in the previous section.

Clause 31 – Determination of main contractor's employment under the main contract

This clause distinguishes between the default of the contractor and the employer. Obviously if the contractor's employment under the main contract is determined by the employer under Standard Form clause 27 (ie, the contractor is at fault) the employment of the subcontractor will also cease. Under clause 31.1 the subcontractor's position in regard to removal of temporary buildings, plant, etc and payment from the contractor is the same as under clause 30.2 (see section 6.4.5), where the subcontractor determines his employment under the subcontract.

On the other hand if the main contractor determines under clause 28 (ie, the employer is at fault) the subcontractor will share pro-rata in payments due under Standard Form clause 28.2.2.1 to 5 (see section 3.6.14). In claims for loss and/or expense on determination under Standard Form clause 28.2.2.6 the subcontractor must supply the contractor with the evidence necessary to establish his claim. It should be noted that if an interim certificate is issued before the contractor determines under clause 28, the amount due therein to the subcontractor is not subject to the pro-rata sharing of payments on determination referred to above (ie, the contractor would have to pay in full).

Clause 32 – Fair wages

This is much lengthier than clause 22 of the 'green' form. It is, in fact, the fair wages clause 19A (suitably amended) which appears in the local authorities edition of the 1980 Standard Form (see section 3.6.8) and operates where 19A applies.

7

OTHER JOINT CONTRACTS TRIBUNAL FORMS

The widening activities of the JCT are welcome because an increasing number of conditions of contract acquire improved status, greater authority and wider acceptance.

In addition to the Standard Forms of Contract the JCT issues the following forms:

(1) Fixed Fee Form
(2) Minor Works Form
(3) Renovation Grant Forms
(4) Standard Form of Tender for Nominated Suppliers under clause 36 of the 1980 edition of the Standard Form.

A significant addition to the forms published by the JCT is the Design and Build Form which made its appearance in 1980 and is reviewed later in this chapter. The new form of contract is entitled 'With Contractor's Design', but there is also another form which incorporates a schedule setting out modifications to clauses in the 1980 Standard Form with Quantities (local authorities edition) entitled 'For use with Quantities and Contractor's Proposals'. The purposes of these two publications are fully explained in the last section of this chapter (7.5) which is devoted to the JCT Design and Build Forms.

7.1 Fixed fee form

It is not always possible at the outset to produce all the drawings and bills of quantities for every building project because of the complex or indeterminable nature of certain jobs.

Carrying out work of this character on a purely daywork basis, ie, prime cost plus percentages for overheads and profit, may be unattractive to the employer since, as prime cost rises, so do the percentage additions. The alternative is to define the extent of the work, which should not be subject to alteration and would be valued on a prime cost basis and to invite contractors to quote a fixed fee to cover overheads and profit, in competition if necessary.

To meet this situation the JCT issued the Fixed Fee Form of contract in 1967 to provide suitable contract conditions where work is carried out under an architect on this basis. It was amended in 1972 (Amendment F1), in 1973 (Amendment F2), and in 1976 (Amendments F3 and F4 – tax deduction scheme) to bring it into line with the Standard Form.

Since the Fixed Fee Form closely follows the Standard Form, the publication of the 1980 edition of the latter will result in the Fixed Fee Form being revised. As this book is being completed this revision has not yet been published but the meaning and intention of the form is not likely to be fundamentally altered when the new edition appears aligned to the 1980 Standard Form. The clause references below are therefore to the pre-1980 Fixed Fee Form.

There are six important schedules to the contract:

(1) definition of prime cost
(2) fixed fee
(3) estimated prime cost, divided between builder's work, nominated subcontractors and suppliers
(4) work to be executed by nominated subcontractors
(5) goods to be supplied by nominated suppliers
(6) items of work to be executed by other persons

The important features of this form are:

(i) The work is clearly defined, preferably in drawings and a specification.
(ii) An estimate is prepared of the prime cost as defined in the First Schedule. This figure is significant in calculating interim payments of the fixed fee.
(iii) The contractor's fixed fee, set out in the Second Schedule, covers overheads, profit and other costs not covered by the prime cost definition. The calculation of this fee, which is not adjustable, is important.
(iv) The Third Schedule contains all the values of the estimated prime cost of builder's work, the nominated subcontractors' work, itemised in the Fourth Schedule, and goods supplied by nominated suppliers set out in the Fifth Schedule.

Many clauses are broadly similar to those in the Standard Form – practical completion, defects, damages, extensions of time, loss and expense, determination, nominated subcontractors and suppliers, insurances and VAT. The main differences derive from the special nature of the arrangements which the Fixed Fee Form is designed to cover.

Clause 3 enables the architect to issue instructions but these must not alter the nature or scope of the works. If he were to do so, then the fixed fee would become inequitable. It is an instruction to alter the scope of the works if the architect requires the contractor to carry out work which a nominated subcontractor as detailed in the Fourth Schedule, or any other person as detailed in the Sixth Schedule, was intended to execute. Where this happens and the contractor agrees to carry out the additional work then a separate or supplemental agreement is called for.

There are special provisions on payment under clauses 26 and 27:

(a) interim certificates issued at the stated period in the appendix, usually at monthly intervals and to be honoured within 14 days, include:

   (i) prime cost of work properly executed, materials properly delivered to site up to not more than 7 days before certificate, and approved off-site materials less payments previously made
   (ii) a deduction for agreed retention money
   (iii) an instalment of the fixed fee being the proportion of the work certified in the certificate under (i), to the estimate of the prime cost of the works set out in the Third Schedule
   (iv) any amounts ascertained by the architect and not previously paid in respect of a loss or expense claim because the regular progress of the works was disturbed under clause 20 or because of compliance with clause 30 (antiquities).

(b) final certificate, to be issued as soon as practicable and before the expiration of three months from the end of Defects Liability Period or of making good defects, if later, and stating:

   (i) the amounts paid under interim certificates and retention released
   (ii) the sum of the prime cost and the fixed fee in Second Schedule adjusted for any additional amounts under clauses 20 and 30 as in (a.iv) above
   (iii) the difference between the sums in (i) and (ii) is the balance due.

In work of this character the provisions of clause 26.3 are very important. These state that: 'The contractor shall keep full and accurate accounts of, and all invoices and records relating to, all payments and work performed for the purpose of this Agreement'. It is essential, therefore, for a contractor to have an efficient costing system and prompt and satisfactory accounting procedures to ensure that expenditure is fully and quickly recovered.

7.2 Minor works form

First published in 1968 by the JCT, this agreement, containing in an abbreviated form certain of the standard conditions, provided a suitable basis for carrying out minor building works or maintenance work under the supervision of an architect for an agreed lump sum. It was not appropriate for work for which bills of quantities had been prepared, the contract basis being specification and drawings only.

This form was originally intended for work up to, say, £10,000 in value but it became very popular as the following 1977/8 annual sales figures show:

Standard Form (all six editions)	– 95,074
Minor Works Form	– 46,750

This may well have been due to the growing complexities of the standard form and the desire to use simpler conditions wherever possible. In the 1980 edition, while not recommending maximum contract value up to which the minor works form may be used, the JCT indicates that it should only be used for new building works or maintenance work without complex services or requirements, and of short duration.

7.2.1 1980 EDITION CHANGES

The JCT revised the form of agreement in 1977 but in 1980 a thorough revision took place with certain significant changes designed to confine the form to minor works as originally envisaged:

(a) There is no provision for bills of quantities although, curiously, there is a provision for the appointment of a quantity surveyor. (There are however more precise provisions for the valuation of variations – see below.)

(b) There is no longer any provision for the nomination of subcontractors or suppliers.

(c) Although the contract is on a lump sum basis, provision is made for the incorporation of an optional clause (4.5) similar to standard form clause 38, other than for works of limited duration.

The opportunity has also been taken of regrouping and redrafting the provisions and preparing the new edition in a format similar to the 1980 edition of the Standard Form. To keep the agreement as short as possible three provisions have been taken into a supplementary memorandum:

Part A	– Contribution, levy and tax fluctuations (ie, based on standard form clause 38)
Part B	– Value added tax (formerly clause 16)
Part C	– Finance (No 2) Act 1975 – Statutory Tax Deduction Scheme (formerly clause 17)

7.2.2 CONTRACT PROVISIONS

The heading to the conditions emphasises that the form is not appropriate where bills have been prepared, the employer wishes to nominate or full fluctuations are proposed because of the length of the contract.

In the recitals there is now a reference to and definition of 'Contract documents' – the conditions together with the numbered drawings and/or a specification; and/or schedules showing and describing the works and attached to the agreement.

In the agreement the contract sum is specifically stated as being exclusive of VAT, the recitals referring to the pricing of the specification or schedules, or the provision of a schedule of rates by the contractor.

By regrouping the provisions there are now eight clauses as set out in the contents page (with three provisions in the Supplementary Memorandum referred to above). Decimal numbering has been introduced.

Since this form is now widely used it is proposed to review the conditions in some detail commenting specially on significant new provisions.

Intentions of the parties

Clause 1.1 is new, being based on clause 2.1 of the Standard Form, ie, the contractor is to carry out the work using materials and workmanship of the quality and standards specified and to the reasonable satisfaction of the architect where these are a matter for his opinion. The contractor must also carry out and complete the works with due diligence and to the reasonable satisfaction of the architect who must issue all necessary information and certificates and confirm instructions in writing.

Execution of the work

The provisions are now grouped in clause 2. Commencement and completion dates are set out with provisions for extensions of time and damages for non-completion at the weekly rate inserted. The architect must certify the date of practical completion, from which date the defects period will run – normally three months unless otherwise stated. The architect certifies on completion of making good defects (see also clause 4.4).

Control of the work

Clause 3 requires the contractor to obtain prior written consent to assignment as well as subletting. Architect's written instructions must be carried out at once, and if within seven days of written notice requiring compliance the contractor still fails to do so, the employer can arrange for someone else to do the work and recover from the contractor. Oral instructions are to be confirmed in writing.

For small work it may seem onerous to require the contractor to keep on the site a competent person-in-charge. A saving provision may be 'at all reasonable times'. An employee may be excluded from the works on a 'reasonable' architect's instruction.

This clause (3.6) provides for variations (additions, omissions, 'other changes', alterations in order or period of execution of the works). Valuation is to be on a fair and reasonable basis using where relevant, prices in the priced specification or schedule of rates where provided or on the basis of agreement prior to execution.

Subclause 7 limits instructions to the expenditure of provisional sums to be valued as in subclause 6 above, nomination being specifically excluded.

Payment

Clause 4 brings together the payment provisions, subclause 1 stipulating that the conditions take precedence over the other contract documents, any inconsistency between drawings, specifications or schedules to be corrected and changes where appropriate treated as a variation under clause 3.6.

Payments are in three categories by means of an architect's certificate to be honoured within 14 days of its date.

Clause 4.2: Progress payments on value of work done, including variations, expenditure of provisional sums and materials on site, if requested by the contractor, at intervals of not less than four weeks – less retention at 5 per cent or other agreed rate as inserted.

Clause 4.3: Penultimate certificate within 14 days of the certified date of practical completion for value of work done as far as ascertainable, less retention and progress payments.

Clause 4.4: Final certificate within 28 days of receipt of documentation and subject to a certificate on making good defects under clause 2.5. In the conditions a period from practical completion is inserted (normally three months) within which the contractor must supply all the documentation reasonably necessary to enable the architect to certify the final amount due.

On possible fluctuations there are alternative provisions – clause 4.5 is deleted if on a short contract period there is a fixed price (ie, as stipulated in alternative clause 4.6 no account is to be taken of any cost increases for labour, materials, plant, etc).

If clause 4.6 is deleted and clause 4.5 stands, the fluctuations for contribution, levy and tax fluctuations are as set out in Part A of the Supplementary Memorandum. The percentage addition to be inserted in clause 4.5.

Statutory obligations

Clause 5 covers such diverse matters as VAT, fair wages, etc.

Subclause 1 requires the contractor to comply with all statutory notices and requirements and pay all fees legally recoverable from him. This clause has been extended to include provisions similar to clause 6.1 of the Standard Form on divergences between statutory requirements and contract documents, instructions, etc.

Two subclauses refer to the Supplementary Memorandum:

5.2 – Part B:VAT
5.3 – Part C:Statutory tax deduction scheme where the employer was or becomes a 'contractor'.

Subclauses 5.4 on fair wages and 5.5 on prevention of corruption apply only where the employer is a local authority.

Injury, damage and insurance

The contractor's liability for injury to or death of persons (6.1) or for damage to property (6.2) arising out of or in the course of the execution of the works is in similar terms to the earlier edition of the agreement. There is a liability to indemnify the employer in respect of personal injury to or the death of any person arising from the works unless due to the employer's negligence, the contractor to maintain, and require any subcontractor to maintain, the necessary insurance. The liability to indemnify and insure in respect of damage to real or personal property extends only to negligence by the contractor or any subcontractor or persons for whom they are responsible.

A new clause 6.3 has alternatives A and B covering insurance against fire, etc – 'A' relating to new works where the contractor is to insure in joint names only the monies received to be available for reinstatement, etc, and 'B' relating to existing structures where the risk of damage to the works, materials, etc is that of the employer who must insure accordingly, furnishing proof to the contractor when reasonably required.

(Similarity to the provisions of clauses 20, 21 and 22 of the Standard Form 1980 edition should be noted.)

Determination

Clause 7 provides for determination by the employer (7.1) and by the contractor (7.2)

The employer may determine where the contractor without reasonable cause wholly suspends the works or fails to proceed diligently or becomes bankrupt, etc. There is a new provision that on determination the contractor must give up possession of the site immediately, the employer not being bound to make any further payment to the contractor until after completion of the works.

In the case of determination by the contractor there must be continuation of the employer's default for seven days after a specifying notice on (1), (2) and (3) below, before notice to determine can be given. The relevant employer's defaults are:

(1) failure to make progress payments when due
(2) interference or obstruction of the works or failure to make premises available
(3) continuous suspension for at least a month
(4) bankruptcy or liquidation

On determination the employer must pay the balance due to the contractor for work done, materials on site, removal of plant, etc.

The procedures (eg, use of registered post, etc) are set out, with the normal provisos that notice must not be given unreasonably or vexatiously and that the right of determination is without prejudice to the other rights and remedies of the parties.

Arbitration

The arbitration provision is in simple terms with the arbitrator to be appointed by the President or a Vice President of the RIBA or the RICS if the parties fail to agree on a suitable person (article 4).

Supplementary memorandum

As already explained this is supplemental to the main agreement and would apply in whole or in part according to the wishes of the parties or the relevant facts – ie:

Part A – fluctuations to apply if clause 4.5 is not deleted
Part B – value added tax as referred to in clause 5.2
Part C – statutory tax deduction scheme provisions to apply where the employer was at the date of tender, or became at any time

during the currency of the contract, a 'contractor' under the scheme (clause 5.3)

The JCT's intention is that the memorandum should be kept for reference and need not be bound in with the executed agreement.

The three parts do not call for special comment since they are similar as far as appropriate to the relevant provisions in the Standard Form 1980 edition (see Chapters 3 and 4) as indicated below.

Part A: This follows the intentions of Standard Form clause 38, the only significant difference deriving from the fact that there are now no nominations under the Minor Works Agreement. The percentage addition is inserted in clause 4.5 of the agreement.

Part B: The value added tax provisions are in almost identical terms to the Supplemental Agreement of the Standard Form.

Part C: The terms of clause 31 of the Standard Form are virtually repeated here.

The agreement is suitable for use in both private and local authority contracts but it is not for use in Scotland.

7.3 Renovation grant forms

The further stimulus to renovation grant work by the Housing Acts 1969 and 1974 resulted in the JCT publishing two forms of contract for this type of work. One is for lump sum contracts where an architect is employed with specification and drawings prepared, and the other on the basis of an estimate of cost by the contractor where no architect is employed. The forms are not appropriate for use in Scotland. The revised edition was issued in December 1975.

(i) The 'architect form' is similar to the pre 1980 Minor Works Form in its provisions for private work but with special references to the renovation grant aspect:
   (a) The agreement sets out the expenditure approved for grant purposes and the amount of the grant.
   (b) The architect must ensure that the grant aided part of the work is carried out and completed to secure local authority approval (clause 1).
   (c) The architect must advise the contractor whether changes in the work will qualify for grant and the contractor is to be reimbursed for estimating for extra work if it is *not* carried out (clause 2.ii).
   (d) Times must be stipulated for the premises to be available to the contractor (clause 6.iii) but with no provision for liquidated damages for delays.
   (e) Evidence of fire, etc insurance should be produced to the contractor by the owner (clause 8.iii).

(f) The building owner must on signing the contract authorise the payment of the grant direct to the contractor where this course is agreed. (clause 10.iv.a). A form of authority is included as an appendix to the contract.

(g) Since the grant may not cover the total cost of the work, certificates must indicate the proportion of grant to be paid direct by the local authority to whom a copy of the certificate must be sent. Within 14 days of the date of the certificate the building owner must pay the balance certified, direct to the contractor. Any shortfall in the grant payment is a debt due from the building owner to the contractor (clause 10.iv.b, c, d).

(ii) The agreement where no architect is appointed sets out the approved expenses for grant work and the amount of the grant. This form is simpler but has similar clauses for VAT, determination and arbitration.

The contractor is required to carry out and complete the works and to provide the data for grant approval and payment (clause I). The conditions cover times for access to the premises (clause 2), provisions for interim payments if the period for the work exceeds 8 weeks (clause 3) and for prompt, final payment of the balance due under the contract (clause 5). If the contractor agrees to carry out any extra work, the cost of the estimate and data for the grant claim will fall on the employer if the work is not approved and a grant refused (clause 7.iii). An authorisation from the building owner to the local authority to pay the grant direct to the contractor under clause 6 forms an appendix to the contract. This takes care of a point which has been a frequent source of concern to builders on this type of work.

7.4 Nominated Suppliers Tender Form

A new 'Standard Form of Tender for Nominated Suppliers' has been produced by the JCT for use in conjunction with clause 36 of the 1980 edition of the Standard Form. Although this tender form is not mandatory, it can be used with considerable advantage in the nomination of suppliers and its use is recommended by the Tribunal. The form of tender has three schedules:

(1) description, etc of goods, main contract details (whether under hand or under seal), completion date, operation of clause 25, and defects liability period; price, delivery dates and access; fluctuation provisions may be set out by the architect or submitted by the supplier
(2) the text of clause 36.3 to 5 of the Standard Form (for details see section 3.6.6)
(3) the warranty by the supplier to which he and the employer are parties.

The architect inviting tenders fills in details about the job, the employer, main contractor, description of goods, main contract completion date and defects liability period, delivery dates etc. The tender form is then completed by the supplier and returned. On being nominated, and receiving an acceptance order from the contractor with whom delivery dates have been agreed, the supplier then:

(a) Confirms that he will be under contract to the main contractor to supply the goods described in Schedule 1 as a nominated supplier under the terms of clause 36.3 to 5 (set out in Schedule 2).
(b) Agrees that the tender will remain open for the period specified from the date of tender (with a right to withdraw within 14 days of the main contractor being subsequently named).
(c) Declares that the warranty agreement (Schedule 3) will be accepted by the supplier once he receives the nomination, subject to his right to withdraw – (b) above. No provision in the warranty agreement will take effect unless and until the supplier has received a copy of the nomination instruction, the main contractor's order accepting the tender, and a copy of the warranty/ agreement signed by the employer.

In the warranty agreement the supplier on being nominated warrants to the employer, subject to the tender conditions, that he has exercised all reasonable skill and care in design, selection of materials and any performance specification; that he will supply such information as the architect or contractor may reasonably require and also keep to the delivery dates agreed with the contractor so that the contractor will have no grounds for claims for consequential extensions of time (clause 25) or direct loss and/or expense (clause 26); but that if claims do arise the employer will be indemnified against inability to recover damages (which are noted in the warranty) because of extensions of time, or claims made under clause 26. Nothing in the tender will exclude or limit liability for any breach of this warranty. It is interesting to compare these provisions with NSC/2 reviewed in section 6.2.1.

Where the joinder provision of the main contract applies, if a dispute arises and the issue is substantially the same as one which is already before an arbitrator under the Standard Form and notice of dispute has been given by either party, then the matter is referred to that arbitrator whose award will be binding on the employer and supplier (see also section 3.10 on arbitration).

7.5 JCT Design and Build Forms

It may at first sight be strange that the Joint Contracts Tribunal, on which the professions are fully represented, should undertake the work of preparing a form of contract 'with contractor's design'.

Certainly 'design and build' is a growth sector in the industry and the Tribunal apparently received representations that, particularly for the purpose of public authority housing, there was a need for a form under which the contractor would both design and build to the employer's requirements. There were apparently also projects where there could be a division of functions – the contractor to design a part with the remainder on the traditional basis of architect's design with bills of quantities, etc.

In due course in 1980 the Tribunal published two new forms:

(i) The Standard Form of Contract with Contractor's Design (likely to be known as the JCT Design and Build Form) where the contractor is responsible for both design and construction services. This is based on the principles, as appropriate, of the Standard Form of Building Contract (local authorities edition without quantities) 1980.

(ii) A modification of the Standard Form with Quantities (local authorities edition) 1980 where the contractor designs and builds part and the remainder is on the traditional basis of architect design/builder construct. (This form is entitled the 'Standard Form for use with Quantities and Contractor's Proposals').

The interest of the JCT in preparing these forms is more readily understood when one sees the stress laid on the availability of professional services for the employer in formulating his requirements, appraising tenders, advising on execution and payment for work and issuing instructions, if so authorised.

The two following sections examine the salient features of both forms set against the background of the Standard Form which is fully reviewed in Chapter 3. It should be remembered that the NFBTE published a Design and Build Form in November 1970. This is considered in Chapter 8 since contracts under this form will still be current. Although it is superseded by the JCT form, it is possible it may continue to be used by those who prefer shorter and simpler (if somewhat different) contract provisions.

7.5.1 THE FORM WITH CONTRACTOR'S DESIGN

Basic principles

The JCT Design and Build Contract is based on the 1980 edition of the Standard Form of Building Contract (without quantities) but substantial changes are necessary because of the nature of the arrangements. In reviewing this form reference will be made only to significant differences, the Standard Form clauses where relevant having been dealt with in Chapter 3. Some clauses substantially reproduced in the design and build form required certain essential minor amendments eg, all references to the architect have been changed to the employer; but apart from changes in clause numbers these are basically the same in intention. Other clauses have been specially drafted to reflect the special relationships. This is considered in detail later by means of a comparative schedule but it is important that the basic concepts of the design and build provisions as set out below, are fully appreciated.

(a) Significant changes from standard form

With the contractor responsible for design the following are the new elements in the contract terms:

- The 'Employer's Requirements' state clearly the employer's needs on which the 'Contractor's Proposals' are prepared and set out in drawings, documents, etc.
- The 'Employer's Agent' is nominated by the employer (in Article 3) (eg, an architect) but he is not a party to the contract and merely advises the employer. Under clause 5 he must provide the contractor with copies of the contract documents and under clause 11 of the conditions he must have access to the works and workshops. In addition the employer's agent, by a written notice by employer to contractor, may be nominated to receive or issue instructions and notices under the contract. Apart from this, the employer assumes most of the functions and responsibilities of the architect under the Standard Form.

(b) Relationship of employer's requirements and contractor's proposals

The employer prepares his requirements, with the advice and guidance of an architect or other adviser he may decide to employ. The employer's requirements (along with the contractor's proposals) are referred to in Article 4 and defined in Appendix 3 of the contract which is reviewed below. These requirements must be firm and not tentative and can range from a description of the accommodation required to a full professionally prepared 'Scheme Design'. These must not be added to later except where, with the consent of the contractor under clause 11, the employer's design requirements are changed (see below 7.5.1).

The tendering procedure is a matter for decision by the employer, but if competition is contemplated, it would be from selected firms with the necessary design (and construction) expertise (see section 2.3.2 on 'Two Stage Tendering').

On the basis of the employer's requirements the contractor submits his proposals in drawings and documents to meet fully the employer's requirements, for a lump sum price payable in stages or periodically by valuation. Since there is no provision for priced bills or a schedule of rates, a contract sum analysis must be prepared (there is no specified form) and annexed to the contractor's proposals. This would be used if any unavoidable changes in the employer's requirements required valuation. There is no provision for PC sums for nominated subcontractors and suppliers.

(*Note*: For work subcontracted by the main contractor a suitable subcontract form is being prepared by the NFBTE based on the non-nominated ('blue') subcontract.)

(c) Discrepancies and divergences

It is absolutely fundamental that there should be complete consensus between what the employer requires and what the contractor proposes. The contract devotes considerable attention to any problems arising from discrepancies and divergences which can be summed up as follows:

(i) In recital 2 the employer states that he has examined the contractor's proposals and is satisfied they meet his requirements (subject to the contract conditions). Where the employer has accepted in the contractor's proposals a divergence from his original requirements these should be amended in the employer's requirements before the contract is signed.

(ii) Clause 2 would appear to cover all eventualities. Under subclause 1 the contractor is to carry out and complete the works (eg, design, selection of materials, workmanship) in accordance with the employer's requirements, the contractor's proposals, the articles of agreement and the conditions which take precedence over the other documents.

(iii) Any discrepancy between employer's requirements or contractor's proposals are to be dealt with under subclause 4 as follows, with immediate written notice to be given to the other on a discrepancy being discovered:

(a) discrepancies within employer's requirements – contractor's proposals to prevail without added cost.
(b) discrepancies within contractor's proposals – contractor to inform employer in writing of proposed amendment to remove discrepancy at no cost to the employer, who must notify his decision in writing
(c) in a divergence between drawings, description, quantities, measurement, etc in employer's requirements and contractor's proposals, the error is to be corrected on fair and reasonable agreed terms.

(iv) An employer's instruction which changes employer's requirements because of a divergence between these and the definition of the site boundary, which he must give under clause 7, falls to be valued under clause 12.

(v) Immediate written notice must be given by either party specifying any divergence noted between statutory requirements and the employer's requirements or contractor's proposals (clause 6). The contractor with the employer's consent to his proposed amendments, will amend the design accordingly at his own cost, the employer to note it in the contract documents. Any cost resulting from a change in statutory requirements after the date of tender are to be paid for by an adjustment to the contract sum.

(d) Design liability

Clause 2.5 makes it clear that the contractor's responsibility to the employer for any inadequacy of design work included in the contractor's proposals or to be prepared in accordance with the employer's requirements will be the same as that of an architect or other professional designer under a design contract. (Contractors would insure for this as far as possible under a professional indemnity policy.) Where a dwelling is involved this reference includes liability under the Defective Premises Act 1972. Where the act does not apply the contractor's design liability for loss of use or profit or consequential loss is limited to the amount, if any, set out in Appendix 1.

Note: Under the Defective Premises Act 1972 a statutory duty is placed on any person taking on work for the provision or improvement of a dwelling. This duty, which is owed to any person ordering a dwelling or to any other person who acquires an interest in it, is to see that work is done in a workmanlike or professional manner with proper materials so that when it is completed the dwelling is fit for habitation. Where work is carried out on the instructions of another, the duty is discharged by compliance, except where there is failure to discharge a duty to warn of defects. Action for breach must be brought within six years of the completion of the dwelling.

Contract provisions

(a) Comparative review

Having dealt with fundamental points of principle we can now look generally at the contract conditions.

Certain clauses of the Standard Form have been substantially amended to meet the design and build concept. Since this Design and Build Form is an entirely new one a comparative schedule is set out below showing the relationship of various clauses to those in the 1980 Standard Form local authorities edition on which it is based.

In certain instances the clauses are virtually identical subject to any minor amendments due to the different relationships, but because of this relationship some clauses are substantially varied. In the former the details can be studied in Chapter 3, full references being given for this purpose. In the latter, the clauses marked with an asterisk are given special consideration in the following section headed 'Special Contract Provisions'.

Design and Build Form clause no	**Subject matter**	**Standard Form 1980 clause no**	**Chapter 3 reference**
1	Interpretations, definitions etc	1	–
2	*Contractor's obligations, etc	–	–
3	Contract sum – additions etc	3	3.8.2
4	Employer's instructions	4	3.5.2
5	*Contract documents	–	–
6	*Statutory obligations, etc	–	–
7	Site boundaries	–	–
8	*Materials, goods and workmanship	8	3.6.3
9	Copyright, etc	9	3.9.6
10	Person-in-charge	10	3.9.1
11	Access to works	11	3.5.2
12	*Changes and provisional sums	–	–
13	Contract sum	14	–
14	VAT-Supplemental Agreement	15	3.8.4
15	*Materials and goods unfixed or off site	16	3.6.3
16	*Practical completion and defects liability	17	3.6.12
17	Partial possession by employer	18	3.6.10
18	Assignment or subletting	19	3.6.7
19	Fair wages	19A	3.6.8
20	Injury to persons and property and employer's indemnity	20	3.7.1
21	Insurance against injury to persons and property	21	3.7.2

Design and Build Form clause no	**Subject matter**	**Standard Form 1980 clause no**	**Chapter 3 reference**
22A, B, C	Fire, etc, insurance (clause 22 perils)	22A, B, C	3.7.2
23	Possession, completion and postponement	23	3.6.1
24	Damages for non-completion	24	3.6.1
25	Extensions of time	25	3.6.9
26	*Loss and/or expense – disturbance	26	3.6.13
27	Determination by employer	27	3.6.14
28	Determination by contractor	28	3.6.14
29	Works by employer or persons employed or engaged by employer	29	3.9.7
30	*Payment	–	–
31	Finance (No 2) Act 1975 – statutory tax deduction scheme	31	3.8.5
32	Outbreaks of hostilities	32	3.9.3
33	War damage	33	3.9.4
34	Antiquities	34	3.9.5
35	*Fluctuations (alternatives)	37	See Chapter 4
36	*Fluctuations ('limited')	38	
37	*Fluctuations ('traditional')	39	
38	*Fluctuations (formula)	40	
Article 5	*Arbitration	Article 5	3.10
Appendix 1	Details on insurance, programme dates, retention, etc		
Appendix 2	*Interim payments:		
	A – Stage payments	–	–
	B – Periodic payments (para 3)	30.3	3.6.3
Appendix 3	Employer's requirements contractor's proposals	—	—

(b) Specific contract provisions

Clauses 2 and 5 – Contractor's obligations and contract documents: As indicated in the preceding section the contractor under clause 2 undertakes to carry out and complete the works referred to in the employer's requirements and contractor's proposals, the articles of agreement and the contract conditions. Discrepancies and divergences are dealt with as explained in 7.5.1 above, and the provisions of clause 2.5 on design liability are also set out.

Clause 5 requires that the employer's requirements and the contractor's proposals (including the contract sum analysis) will remain in the custody of the employer to be available for inspection by the contractor, two copies of each to be furnished to the contractor without charge. The contractor is also entitled to a certified copy of the articles and the conditions and he in turn must provide free of charge two copies of drawings, specifications etc. There must also be available on site for the employer's agent one copy of all these documents.

Before the commencement of the defects liability period the employer must be provided, free of charge for his retention and use, with the drawings and details of the works as built and their maintenance and operation including installations. All the documents referred to will only be used for the purposes of the contract. There is also a provision in clause 27, on determination by the employer, for the contractor to supply him free of charge with all documents including drawings, details, etc related to the works up to the date of determination.

Clause 6 – Statutory notices: Clause 6 is specially drawn to place on the contractor full responsibility for paying fees and charges and for complying with and giving notice under any Act, regulations or byelaw, including development control requirements (including planning permission, byelaw approvals, etc). All approvals received are to be passed to the employer. The provision on the effect of statutory requirements on design, etc is referred to above.

In clause 6.1.3 there is cover for emergency compliance, the employer to be informed at once. All fees and charges (including rates) are to be included in the contract sum and met by the contractor, unless covered by a provisional sum in the employer's requirements.

Clause 8 and 15 – Materials: The provisions of an adapted clause 8 are similar to those of the Standard Form, the employer taking the place of the architect and approving the substitution of alternative materials. Clause 15 on unfixed and off-site materials has the same intention as clause 16.1 of the Standard Form. Appendix 2B.3 sets out the terms of Standard Form clause 30.3 for the inclusion of off-site materials in interim payments at the employer's discretion.

Clause 12 – Variations: This clause has the title of 'Changes and Provisional Sums' because it visualises that, although it is the intention that the initial employer's requirements should be comprehensive and complete, some changes may be required by circumstances, including statutory requirements under clause 6 (above). Clause 12 stipulates that where an instruction by an employer alters or modifies the contractor's design, his consent is required. Changes in employer's requirements may extend to the design, quality and quantity of the works including any addition, omission, alteration, removal or substitution of work and materials. The employer must also issue instructions under clause 12 on the expenditure of provisional sums included in his requirements.

The rules for valuing these changes, including the expenditure of provisional sums, are set out in clause 12.5 in terms very similar to clause 13.5 of the Standard Form. Relevant prices are, of course, those in the contract sum analysis. The resultant adjustments are to be made in the contract sum and included in interim payments.

Direct loss and/or expense which would not be otherwise reimbursed under a clause 12 valuation in respect of a change in the employer's requirements, or the execution of work under a provisional sum, must be added to the contract sum and included in interim payments, provided the contractor has made a written application within reasonable time of the loss being incurred.

Clause 16 – Defects: The date of practical completion is fixed by a written statement by the employer. The defects liability period in the Appendix (usually six months) runs from this date, the employer to deliver for prompt attention by the contractor the schedule of defects within 14 days of its expiration after which no further instruction can be issued. The employer can of course ask for defects to be made good during this defects liability period. When all defects are made good, the employer must issue a notice accordingly. (cf, clause 17 Standard Form). The removal or rectification of work, materials, etc not in accordance with the contract may be required by the employer under clause 8.4.

Clause 26 – Loss and/or expense – disturbance: As in the Standard Form clause 26 there are grounds for loss and/or expense claims for matters materially affecting progress occasioned by the employer eg, inspection of work, delays caused by persons employed direct by employer, or postponement or failure to give promised ingress to or egress from the site. There is also provision where there is delay in the receipt of development control permissions the contractor to do all he can to avoid delays. (These could also give grounds for an extension of time under clause 25; and determination by the contractor under clause 28.1.2.8.)

Clause 30 – Payment: Appendix 2 sets out the arrangements for payment. Interim payments are made by stages or by periodic monthly payments up to practical completion and thereafter as amounts are due. These payments are to be made within 14 days of each application by the contractor, in accordance with the payment terms. Applications are to be supported by an account of adjustments such as changes in employer's requirements, expenditure of provisional sums, loss and/or expense claims, fluctuations, etc. For periodic payments there must also be an itemised statement of work, materials and goods. Retention percentages and principles are in line with clause 30 of the Standard Form.

The final account with documents must be sent to the employer within 3 months of practical completion. A time limit is placed on the right of the employer to dispute the final account – one month after the end of defects liability period (or making good defects), or four months of its submission, whichever is the later. Thereafter the account is conclusive as to the balance due. The agreed final account will be conclusive evidence that the employer is satisfied with the quality of materials and standards of workmanship as stated in his requirements.

If amounts payable cannot be agreed with the employer, arbitration is available under article 5.

Clauses 35 to 38 – Fluctuations: These are provided for in terms similar in principle to clauses 37 to 40 of the Standard Form. (Traditional 'fluctuations' or price adjustment formula.) It has been found possible to prepare formula rules for use with this form and these have been published by the JCT together with revised formula rules for use with the 1980 editions of the Standard Form and the JCT Nominated Subcontract Form (NSC/4). The formula principles are fully explained in Chapter 4.

7.5.2 CONTRACTOR'S DESIGNED PORTION – CONTRACT ADDENDUM

The potential difficulties of a contractor constructing the whole and designing only part of a project are reflected in the modifications to the Standard Form with quantities (local authorities edition) which the JCT has published in an endeavour to provide satisfactory composite contract conditions for this hybrid situation.

One wonders in fact how often this sort of arrangement is made and with what results.

Principles and procedures

The concept is that the Standard Form covers the architect designed/builder constructed section in the usual way with the architect and quantity surveyor preparing drawings, specifications and bills of quantities. The portion designed by the contractor is the subject of a separate provision, the conditions being an adaptation of the Standard Form, as set out in a schedule of modifications referred to in Article 1 of the agreement. This form is known as the Standard Form of Building Contract, local authorities 1980 edition 'for use with Quantities and Contractor's Proposals'.

In principle the relationships and the documents for the contractor's designed portion are similar to those in the Design and Build Form, already reviewed, but as will be seen the architect plays a more significant part.

The procedure is that for the contractor's designed portion' the necessary documents known as the employer's requirement's are prepared by his architect. The contractor in response submits the contractor's proposals and an analysis of the price related to the designed portion.

The drawings, priced bills, together with the employer's requirements and contractor's proposals and price analysis are signed and identified in a schedule to the articles of agreement; all these together with the conditions are to be the contract documents.

Two aspects of this arrangement call for comment. Firstly the architect acts contractually for the employer in the contractor's designed portion as in the usual Standard Form arrangements, whereas in the Design and Build Form the employer exercises most of the functions. Secondly it is necessary to regulate carefully the relationship between architect and designer-builder including integration and interpretation of design and to allow for any architect's observations on possible defects in the builder's design.

Contract provisions

(a) Design etc

The contractor agrees, subject to the standard conditions as modified, to carry out and complete the work shown in the contract drawings and described in the bills, the employer's requirements and contractor's proposals and to complete his design in accordance with any architect's directions on integration with the design of the works. If the contractor considers that compliance with any direction will injuriously affect the efficiency of his design (or one prepared on his behalf) he must specify this in writing within seven days of the direction, which shall not take effect until confirmed.

The contractor's responsibility for design will be that of an architect or other appropriate professional designer under a separate design contract. This extends to liability under the Defective Premises Act 1972 where it applies. Where it does not apply the liability of the contractor for design liability for loss, damage, etc will be limited to the amount, if any, stated in the schedule (see 7.5.1 above). A contractor will not be relieved of his design obligations whether or not the architect gives notice of anything which appears to him to be a defect in design.

(b) Discrepancies

On finding divergences between employer's requirements and contractor's proposals (and any amplifying details) or discrepancies within these proposals, immediate written notice must be given to the other. The contractor must then inform the architect of his proposals to remedy (at his own cost) and the architect will then instruct accordingly on the contractor's designed portion.

The architect must on request be provided free of charge with two copies of all drawings, details, etc in amplification of the contractor's proposals. The contractor must allow 14 days from delivery before commencing work so detailed.

An error in the contract documents consisting of a divergence between drawings, details, etc in the employer's requirements and contractor's proposals must be corrected on such terms as may be considered fair and reasonable.

In regard to statutory requirements extra cost resulting from any changes after the tender date affecting the contractor's designed portion will be added to the contract sum, but the contractor must meet the cost of any divergences between existing statutory requirements and design, the architect to instruct on proposals from the contractor to deal with the situation. There is also provision for emergency action, the architect to be immediately informed.

(c) Miscellaneous

There are provisions similar to those in the Design and Build Form for the supply of drawings, maintenance data, etc before the commencement of the defects liability period. There is no provision for nominated subcontractors and suppliers in the contractor's designed portion.

The architect's consent is necessary for the substitution of materials described in the employer's requirements or, failing that, in the contractor's proposals. Where the architect's approval is required, materials and workmanship must be to his reasonable satisfaction.

There is a reversal of roles in regard to extensions of time and loss and/or expense claims. These provisions will not have effect where delays are due to an error, divergence, omission or discrepancy in the contractor's proposals or if in relation to the contractor's designed portion there is a failure to furnish drawings, etc to the architect in due time, or to supply details for which the architect has applied in writing in reasonable time.

Where variations occur as a result of a change in employer's requirements these will be valued under clause 13 by reference to the contractor's price analysis.

8

MISCELLANEOUS FORMS AND PUBLICATIONS

Several of the main bodies concerned with building contract conditions have published either separately or jointly a variety of other forms and agreements.

8.1 NFBTE Design and Build Form

An owner of land on which he wants to erect a building, and who chooses to have a contractor able to prepare designs, plans and a specification to do the work, has been able to use the NFBTE Design and Build Form of Contract for his 'package deal'.

This form was first published in 1970 but the JCT Design and Build Form, which appeared in 1980 and is reviewed in Chapter 7, supersedes it. However since work is still being carried out under the NFBTE form, and it may continue to be used to a limited extent, it is considered here.

The contractor's responsibilities under this form are wider than usual, since in addition to the design element he must obtain the prescribed town planning and other consents, other than an Industrial Development Certificate now no longer generally required, before starting work. There is a warranty that all reasonable professional skill and care has been exercised in preparing plans, etc (clause 2), and it is implied that the design, materials, etc, will be fit for the prescribed purpose, although suitable substitutes for materials are allowed under clause 1.2.

The contractor is responsible at a professional level for the fitness of the specification and design for the purpose stated (clause 2) and should, where possible, take out the necessary professional indemnity insurance. Copyright in drawings, etc, remains with the contractor under clause 21. Insurances are needed under clause 9 for the same cover as under clause 19 of the pre-1980 Standard Form, and under clause 9.I.c for public liability joint-name cover under a provisional sum somewhat similar to clause 19.2.a of the pre-1980 Standard Form.

The contractor must indemnify the employer against injury to persons and property and is responsible for the relevant employers and public liability insurances (clauses 8 and 9) and for fire, etc, insurance on new works, the employer being responsible for insuring work on existing buildings (clause 10).

The contract sum which includes design, etc, fees, is to be paid at agreed predetermined stages, and the last payment is to be made when, under clause 6, the defects have been made good at the end of the 6 months liability period (clause 18). When extras or alterations are due to variations ordered by the employer (clause 4), or because of statutory requirements (clause 3), these are separately agreed and paid for to avoid adjustments of the predetermined amounts due under clause 18. Adjustments for PC or provisional sums are covered by clause 5. VAT procedures and payments are dealt with in clause 18A.

There are dates for possession and completion (clause 11), for damages for non-completion (clause 12), and any extensions of time (clause 14) not mutually agreed are to be referred to an arbitrator (clause 22).

Clauses 15 and 16 provide for determination by the contractor or on his bankruptcy but there is no provision for determination by the employer. Clause 19 covers extensive fluctuations if agreed, and clause 13 allows for sectional completion.

Since design and build contracts vary, the terms of this contract should be carefully scrutinised before it is used to ensure that it is entirely relevant. There is no standard form of sub-contract published for use with this form, and it should be noted that it is not suitable for use in Scotland.

8.2 Labour-only subcontract form

This form, first published by the NFBTE in 1964, sets out the conditions for subletting work 'labour only', and is a contract between contractor and the subcontractor. The current edition is 4.76.

It is in the form of an order in triplicate describing the work and setting out prices which are stipulated as VAT exclusive. The general conditions are supplemented by special conditions relating to the particular contract (eg, dates for commencement and completion, defects, retention, services and plant provided by the contractor). The sub-contractor signs the order accepting the conditions and returns a copy to the main contractor.

The general conditions cover variations (clause A) and the defects liability period which is 6 months unless specially provided otherwise (clause B). Employers and public liability insurances are covered in clause C, and the subcontractor's responsibility as employer regarding wage rates and other NJC conditions are to be found in clause D. PAYE is in clause E and National Insurance, etc, in clause F which also stipulates that there will be no adjustment for labour, etc fluctuations unless provided in the special conditions. Payments under clause G provide for weekly instalments and retention money, half of which is released on completion and half on the expiry of the defects liability period.

There are provisions sufficiently widely drawn in clause G.1 to cover the 1972 and 1977 subcontractor tax deduction schemes entitling the contractor to make the statutory deductions from payments to subcontractors who do not hold valid tax certificates (see Chapter 3). The contractor is also entitled to recover from sums due, for any liability arising from the failure by the subcontractor to comply with National Insurance, etc regulations or to make payments to the CITB in respect of the subcontractor's employment.

Where the subcontractor holds a valid tax certificate he must produce it for inspection as stipulated, notify the contractor immediately of its cancellation or withdrawal and comply with the scheme in regard to receipts, the contractor having the right to withhold payment if receipts are not received within seven days of payment (G.2). The contractor must be indemnified against any tax deduction liability where he relies on a certificate which proves to be invalid (G.3).

The VAT clause (G.4) makes alternative provision for (i) authenticated receipts, with a right to withhold further payment if not received within seven days of payment or (ii) a self billing document prepared by the contractor and sent to the subcontractor for completion.

Other miscellaneous provisions relate to loss or damage (clause J), safety (clause L), and determination of contract on seven days notice, in the circumstances prescribed (clause M).

This form is relevant to *bona fide* labour only sub-contractors under the Working Rule Agreement but not to self-employed men. Despite the reluctance of some sub-contractors to sign a contract, its use can avoid difficulties.

8.3 Nominated suppliers

This section relates to the position under the pre-1980 edition of the Standard Form where a nominated supplier might seek to impose conditions of sale at variance with those in clause 28. This could cause trouble and deprive the contractor of his rights in regard to nominated suppliers. The proper course for the contractor to take was to refuse the nomination and ask the architect for instructions.

To prevent this situation arising a Form of Tender was available under the sanction of the RIBA, RICS and NFBTE and prepared after discussion with the National Federation of Building Material Producers. This was to accompany an enquiry by the architect, the tender being addressed to the main contractor.

The main features (referred to in the present tense) are as follows.

The basis of the tender which remains open for the period specified, is set out – eg, a nominated supply under a PC or provisional sum. The supplier undertakes to provide the goods described in the Second Schedule for a specified sum subject to any alterations authorised by the architect, and to price fluctuations which are to be notified promptly in writing to the contractor as defined in the Third Schedule. The period for commencement of delivery from the date of order and the period to complete are set out, with a saving provision for delays due to weather, fire, strikes, accidents, 'or other causes beyond our control'. This goes beyond the provisos in clause 28.

Defects, which should always be promptly notified, must be made good if they appear within the specified period of time after delivery and reasonable expenses incurred by the main contractor are borne by the supplier eg, incidental labour costs in replacing. The defects are those which could not have been revealed before fixing by examination by the contractor, and which are due solely to defective workmanship or material and not to improper storage, misuse or neglect by the contractor etc.

This contract of sale is between the main contractor and the nominated supplier. The RIBA produced a warranty form to establish a contractual relationship between the employer and a nominated supplier, particularly where a design element applies. Under this, the nominated supplier warrants to the employer that all proper skill and care has been exercised in the design of the goods and the selection of materials, that the goods will correspond with the description including any performance specification, and that delivery dates, etc, will be kept subject to proper time extensions.

The supplier must, if requested, provide a performance bond and insurance cover against breach of warranty. The main contractor, however, is not concerned with design and is not a party to this warranty.

It is interesting to compare these provisions with the new JCT Nominated Suppliers' Tender Form which although not mandatory, is recommended by the JCT for use under clause 36 (Nominated Suppliers) of the 1980 edition of the Standard Form. This sets out sale conditions in conformity with clause 36 and includes a warranty to the employer on design, selection of materials and performance specification (see section 7.4).

8.4 New work (including housebuilding) form

New work, including private housebuilding, is often carried out without an architect being employed.

To establish mutually acceptable conditions for both the employer and builder, in 1953 the Eastern Federation of Building Trades Employers produced a form of contract which is widely used and has been singularly free from dispute. It is still published by the Eastern Builders Federation.

Reduced to as few conditions as possible, the agreement has the usual provisions describing the work and setting out the contract price. The value of variations is to be agreed, if possible, before being carried out, and prices are subject to a rise and fall clause, any increases or decreases so calculated to be increased by 20 per cent to cover overheads etc. There are the usual insurance requirements and provision for determination and arbitration.

On new work, other than housing, there are monthly payments of 95 per cent of the work executed. In the alternative clause for housing there are six stage payments. The retention money amounts to 5 per cent and half of it is released on practical completion and the other half once defects have been made good at the end of the six months defects liability period. There is also a provision for interest on overdue payment at 2 per cent above minimum lending rate.

8.5 Construction of new streets

An agreement ancillary to estate development has been prepared by the Housebuilders Federation of the NFBTE for use under the Highways Act 1959 (sections 40 and 192.3.d).

This sets out conditions under which a housebuilder who is developing his own land can come to an agreement with the local authority to provide private streets on an estate to a required standard. The authority later takes these over as highways maintained at the public expense.

The agreement defines the land and the location of the streets by reference to a plan (clause I). The developer agrees to construct the streets 'with all diligence' in accordance with an annexed specification, to the reasonable satisfaction of the council surveyor who has the right of access and inspection under clause 4, and in compliance with the byelaws (clause 2).

Within six months of buildings being completed, they must be connected to an existing highway (clause 3), and this period is only extended where continuing work might cause undue damage to the streets. Until the streets are taken over the developer must keep them in a good state of repair and must make good defects as required by the surveyor (clause 5).

When all these requirements are met the Council then follows the necessary procedure for taking over the streets. This can be done in sections of one hundred yard lengths or more, once they are certified by the surveyor as satisfactorily completed, and connected to the public highway (clause 6). There is a six months defects liability period (clause 7).

8.6 Conditions of estimate

Builders often estimate for smaller jobs on ill-defined conditions, a situation which can be unsatisfactory to both builder and customer. The simple estimate form devised by the NFBTE remedies this, the 1979 edition being revised to take into account the test of reasonableness under the Unfair Contract Terms Act 1977 (see Appendix H).

It consists of an offer (to be confirmed if not accepted within two months) to carry out work as described in the estimate and any annexed drawings and specification, for the price stated, (exclusive of VAT).

The conditions are fair and brief:

(a) The builder provides all labour, materials, etc to carry out the work. Where the customer specifies materials which the builder considers unsuitable and the customer disregards the builder's written advice, the builder will not be responsible for their suitability or any resultant loss or damage, except where he has been negligent.

(b) PC and provisional sums, and their adjustment including a cash discount up to 5 per cent are clearly set out.

(c) The value of variations should, if possible, be agreed before the work is carried out.

(d) Notice in writing of defects must be given to the builder within three months from completion.

(e) Payment is due immediately on submission of a final account and interim payments are envisaged where the work takes more than a month.

(f) The customer must insure against fire, etc, a copy of the estimate and the conditions to be sent to the insurers and adequate insurance maintained. Cover must include the customer's existing structure and contents and also the works and all unfixed goods and materials properly delivered to site (excluding plants, tools, etc).

(g) Credit is allowed in the price for materials removed, unless otherwise specified.

(h) The completion date may be extended for reasons set out (eg, inclement weather, strikes, variations, etc).

(i) There is a provision for arbitration.

8.7 Daywork – definitions of prime cost

As stated in section 3.6.4, a variation under clause 13 of the 1980 Standard Form which cannot properly be measured and valued should, unless otherwise agreed, be valued on a daywork basis under clause 13.5.4. Work of a jobbing or maintenance character may also be carried out on a daywork basis.

Up to 1963 the RICS and NFBTE published National Schedules of Daywork Charges with figures based on cost surveys. There were widely used but were withdrawn following a hearing by the Restrictive Practices Court. Thereafter daywork charges became a matter for separate agreement between builder and customer.

In 1966 the RICS and NFBTE published two documents defining prime cost – one for daywork under a building contract and one for jobbing and maintenance work. There were some differences between the two definitions but they provided a clear cut and acceptable basis of prime cost to which were to be added percentages for overheads and profits as agreed between the parties.

8.7.1 CONTRACT DAYWORK

With the passage of time certain changes, particularly in labour prime cost, took place, with the result that under the 1966 contract daywork definition, contractors did not recover all labour prime cost in valuations under clause 11(4)(c)(i) of the pre-1980 form. This was particularly so in the case of guaranteed minimum bonus (G.M.B.) introduced on 25 June 1973 by a decision of the NJC. This type of payment was not envisaged when the 1966 definitions were agreed and was not covered by the labour prime cost definition. The GMB was deemed to be covered in the overhead percentage under the heading of 'bonuses' (section 4(v)). Curiously enough, the jobbing daywork definitions were more comprehensive.

The definition was therefore revised and a new edition published on 1 December 1975. This applied to contracts tendered for after that date. It did not affect the valuation of daywork under an existing contract when carried out after 1 December 1975.

The new definition widens the scope of labour prime cost, with correspondingly less to be covered by the overhead percentage. It has six sections briefly providing as follows:

(a) *Application*: For valuation of daywork where provided in the contract (eg, standard form clause 11(4)(c)(1) (now 13.5.4.1); 'green' subcontract form clause 10(b)(3)(A); 'blue' subcontract form clause 12(2)(d)(ii); JCT subcontract form clauses 16.3.4 and 17.4.3. It is not applicable to jobbing work, to prime cost contracts, or to daywork under the Standard Form carried out during the defects liability period which may be the subject of a separate agreement. (It will be appreciated that in the defects liability period daywork may require the special return to the site of labour, plant, supervision, etc.)

(b) *Composition*: Labour, materials and plant, all as defined, make up prime cost. To these are added the percentages for overheads and profits (see (f) below). These percentages are set out in the contract bills in the Standard Form (see 1980 form clause 13.5.4.1) and as appropriate in the Appendix of the pre 1980 subcontract forms ('green': part IIIA; 'blue' part IVB) and in JCT Tender NSC/1 p.1.

(c) *Labour prime cost*: Hourly rates are calculated by dividing the annual prime cost of labour based upon standard working hours as defined, by the number of standard working hours per annum. Prime cost is based on NJC rates, emoluments and expenses for standard working hours payable *when the work is carried out* (or the rates, etc of any other appropriate wage fixing body, eg, plumbing JIB). Prime cost includes standard basic rate, guaranteed minimum bonus (GMB), joint board supplement (JBS) etc, extra payment for skill, discomfort, risk, etc (NWR 3), public holidays, employers' contributions for national insurance (on labour prime cost), holiday stamps, and any statutory levy, etc on employers. Where a supervisor works manually, his trade rate is admissible. (It is useful to agree such a situation in advance.)

(d) *Materials and goods*: Where purchased, these should be invoiced at cost after deducting trade discount but allowing cash discounts up to 5 per cent plus delivery charges. Goods from builder's stock – current market price plus handling charges. VAT is not included if the contractor can treat it as input tax.

(e) *Plant*: Rates as in the contract for mechanical plant and transport and non-mechanised plant specifically brought on to the site for the time employed on daywork. The charge for plant already on site may be based by agreement on the RICS Schedule of Basic Plant Charges (the date of publication should be noted). Non-mechanical hand tools, scaffolding, etc are included in overheads (see (f) below).

(f) *Overheads*: In determining the build-up of prime cost, items included in the overhead percentage must be borne in mind. The following are among the items covered in overheads.

Site staff and supervision and head office charges
Overtime premium unless prior written agreement
Inclement weather payments
Bonuses above guaranteed minimum (in section C)
Apprentices' study time
Subsistence and travelling (NWR 6)
Insurance: Employers and Public Liability: Sickness (or sick pay NWR 9); national insurance (not included in section C)
Tools: allowances, use, repair, sharpening, etc
Scaffolding, protective clothing, safety and welfare, storage
Profit

It should be specially noted that under 1980 Standard Form clause 13.5.4, vouchers for time spent and materials used must be delivered to the architect for verification not later than the end of the following week (there are similar appropriate provisions in the 'green' and 'blue' subcontract forms and in NSC/4 clause 16.3.4 and 17.4.3).

8.7.2 OTHER DAYWORK

(a) *Jobbing work*: The 1966 RICS/NFBTE definition of prime cost for jobbing work is now being revised. As mentioned earlier, the 1966 definitions differed somewhat from the Contract Daywork Definition and a special 'work subcontracted' item was included.

To assist builders to calculate their own percentages for overheads and profits as a prerequisite to agreement with their customers, the NFBTE in the past published 'Builders Overheads and Labour Rates' with examples of items which should be regarded as expenses, thus enabling builders to work out realistic figures for their own businesses.

(b) *Specialist work*: Separate definitions of prime cost for specialist trades are recognised in the main form (clause 13.5.4.2), the nominated ('green') form (clause 10(b)(3)(B) and Appendix Part III A) and the non-nominated ('blue') form (clause 12(2)(d)(ii) and Appendix IV B), and the JCT subcontract form NSC/4 clauses 16.3.4 and 17.4.3 identified on page 1 of NSC/1.

Note: Work carried out on a prime cost plus fixed fee basis is covered by the appropriate JCT form (see section 7.1). Prime cost is defined under the First Schedule of the contract.

8.8 GC/Works/1 subcontract form

In preparing the terms on which subcontracts should be carried out under GC/Works/1 there are three important considerations:

(i) In the case of non-nominated (sublet) work condition 30 sets out the matters which must be included in the subcontract.

(ii) There is no similar requirement for nominated work but in the tender documents sent to subcontractors by the Property Services Agency there are detailed conditions on which the tender is based. These however do not form the terms of the contract between the main contractor and his nominated subcontractor which must be separately formulated.

(iii) The main contractor is fully responsible under condition 38 for the work of the nominated subcontractor (see section 9.6).

A suitable subcontract document must take these points into account and ensure that those conditions in GC/Works/1, which are equally applicable to both main contractor and subcontractor, are fully incorporated. This allows the main contractor, on whom the overall responsibility rests, to pass on to the subcontractor any liability arising from his default, etc in respect of the subcontract works.

Until 1980, when the NFBTE is proposing to publish a set of conditions for this purpose, there was no suitable GC/Works/1 subcontract form.

This new form will be appropriate to both nominated and non-nominated subcontracts, will have the approval of FASS and CASEC and be published with the prior knowledge of the Property Services Agency. Chapter 9, which reviews the provisions of GC/Works/1, should be read in conjunction with this section. The form will cover the essential points arising from GC/Works/1 conditions and the main contractor's relationship with his subcontractor (both nominated and non-nominated). Other provisions however will have some similarity to those in the 'green' (nominated) and 'blue' (non-nominated) subcontract forms reviewed in Chapter 5 to which several references are made below.

To make this book as complete and up-to-date as possible, it is proposed to outline here the basic concepts of the form which is in its final draft stage as this is being written. The actual provisions themselves should be carefully studied once the form is published, but what is written here may assist in understanding the new conditions when they appear.

The form will include relevant provisions in GC/Works/1, relate other GC/Works/1 provisions to the subcontract work, and set out conditions specially applicable to the subcontract itself.

8.8.1 SIMILAR PROVISIONS TO GC/WORKS/1

A substantial number of conditions in GC/Works/1 are applicable to both main and subcontract work. It would be essential that these conditions be incorporated by general reference in the subcontract so that the subcontractor is similarly bound by them in respect of his work. Thus it would be appropriate to stipulate that the subcontractor complies with all reasonable directions of the contractor; that he is entitled to inspect and is deemed to have full knowledge of, and will observe all provisions in, the main contract relating to his work and that he will indemnify the main contractor against any breach by him of the main contract provisions and any act, etc involving the contractor in liability to the authority in respect of any claim for negligence or breach of duty on his part. (See, for example, clauses 3(b)(i)(ii) and

(iii) in both the 'green' and 'blue' forms.)

The subcontractor would not expect to be liable in respect of any act or breach of duty on the part of the authority, the contractor or other subcontractors and there would be no contractual relationship between the subcontractor and the authority or any other subcontractor. The conditions of the main contract, with its rights, duties and obligations would, of course, apply equally to the subcontractor as appropriate.

To ensure that the significance of this is clearly understood, there is likely to be a schedule attached to the subcontract which would virtually translate over 30 relevant main contract conditions into subcontract terms and be read in conjunction with the subcontract. It would cover such matters as contract documents, local authority fees, instructions, setting out, workmanship, excavations, suspension of work, partial possession, subletting, measurement, fair wages and special features of government contracts (passes, secrecy, etc). These are fully explained in section 9.1 to 9.

Although it is desirable that any design element should be eliminated from this contractual relationship any significant design work should desirably be covered by a design indemnity (see section 9.6.2).

The written consent of the contractor would certainly be needed for any assignment or subletting and the main contractor should obtain for the subcontractor any appropriate benefits as far as applicable under the main contract.

There are some GC/Works/1 conditions which have no counterpart in the subcontract – eg, subcontractors and suppliers (31), PC items (38) and provisional sums (39).

8.8.2 PROVISIONS ADAPTED TO THE SUBCONTRACT

In addition to those responsibilities which the main contractor would wish to pass on to the subcontractor in respect of his subcontract work, there are provisions in GC/Works/1 which, suitably adapted, should appear in a subcontract, including those under the following headings:

(i) *Carrying out the work*

Compliance (or failure to comply) with instructions from the SO and directions from the contractor.

Valuation of variations.

Commencement, execution and completion of the work, with provision for remedying defects, extensions of time and claims where regular progress is materially affected. (*Note:* The benefits of the new GC/Works/1 Condition 53 should be extended to the subcontractor (see section 9.7.3).)

Responsibility for subcontract works and materials on site (see clause 5 of the 'green' and 'blue' forms).

(ii) *Indemnities and insurances*

Indemnities would be sought from the subcontractor for injury to persons unless due to the negligence of the authority, the main contractor or any subcontractor.

Insurance would be required for injury to persons and property and also to cover subcontractor's plant equipment, etc because of the special vesting provisions (see section 9.5.8).

(iii) *Payment*
The provisions for payment to the main contractor (see section 9.7) should be reflected in the terms of the subcontract (eg, monthly payment – the authority to notify; cash discount; retention (3%) etc). The subcontractor should be enabled to report to the SO when payment conditions have not been met.

Fluctuation provisions (detailed in the Appendix) would normally follow those in the main contract (ie, condition 11 – traditional; condition 146 – formula).

(iv) *Determination*
Determination by the main contractor or the subcontractor is likely to follow the normal subcontract conditions (eg, 'blue' form clauses 21 and 22).

However where the main contract is determined this can be under one of three conditions in GC/Works/1.

Condition 44 – special powers of the authority
45 – main contractor's default
55 – corruption

Under 44 the subcontractor should benefit from the special provisions available to the main contractor.

Under 45 and 55, provided the subcontractor was not at fault, he should be entitled to be paid in full for the work carried out, for materials and for the cost of removing plant, etc.

(v) *Arbitration*
This would call for special treatment in view of the limits on matters which may be referred to arbitration under GC/Works/1 (see section 9.10). The machinery however would be in accordance with building practice, perhaps with a 'joinder' provision.

8.8.3 STANDARD SUBCONTRACT CONDITIONS
There are certain provisions common to all building subcontracts (ie, 'green', 'blue' and JCT forms) which would certainly be incorporated: ie, set-off and VAT.

Copies of this form will be available from NFBTE once published.

8.9 Other forms

There are three other publications of the NFBTE likely to appear in 1980 which could usefully be referred to in this section.

8.9.1 LOCAL AUTHORITY MAINTENANCE WORK
This will provide conditions of contract for local authority maintenance work carried out by contractors on a measured term contract basis using a schedule of rates, together with a model form of tender.

8.9.2 NON-NOMINATED WORK

A new NFBTE Code of Procedure for letting and management of non-nominated subcontract works is designed to achieve more effective tendering and subcontract management. Checklists enable main and subcontractors to apply recommended procedures on invitations to tender, acceptance, activities before and during construction and after practical completion. There is emphasis on the desirability of using the Standard Forms of subcontract.

8.9.3 MODEL CONDITIONS OF PURCHASE

These conditions will cover other than nominated suppliers for which the JCT has produced the Standard Form of Tender for Nominated Suppliers (see section 7.4).

9

GOVERNMENT BUILDING CONTRACT CONDITIONS

Since the Government is by far the biggest single customer of the industry its conditions of contract call for special examination. Despite attempts to unify building contract conditions the Government still uses its own conditions which apply to both building and civil engineering work.

These conditions are distinctive in that they are prepared by Government draftsmen on what can be regarded as a unilateral basis although the views of the major employers' federations are sought on the provisions or any proposed amendments. The current forms for general building work are:

(i) GC/Works/1 (2nd edition September 1977) (General Conditions of Government Contracts for Building and Civil Engineering Works)
(ii) GC/Works/2 (1st edition April 1974) (General Conditions of Government Contracts for Building and Civil Engineering Minor Works)
(iii) Form C 1501 (January 1974) (General Conditions of Contract for Measured Term Contracts: together with particular clauses to the Contract C 1523)

The Works 1 and 2 forms are for general building work. Form 1 is generally used for contracts over £30,000 in value where for the most part quantities are provided. The Works 2 form applies to smaller jobs usually of less than £30,000 in value, based on specification and drawings. The Measured Term Contract caters mainly for the maintenance of Government establishments over a defined period. There are forms for specialist engineering services let direct which are not within the scope of this book.

The forms are reviewed below and GC/Works/1 is compared with the Standard Form of Building Contract revealing some interesting and important differences particularly on subcontracting, determination and arbitration.

Conditions prepared by one party (the client) are likely to have two features – they may be clearer and less involved than those requiring agreement by five or even ten parties, and they should protect the employer's interests, compatible, one hopes, with a fair deal for the contractor. GC/Works/1 is, in some respects, more clearcut in its provisions than the Standard Form. The employer (the 'authority') has extensive powers, certain provisions in favour of the contractor being exercised on the decisions of the authority (or its agent, the superintending officer (SO)). Several of the decisions are final and conclusive and not referable to arbitration. However, as in insurance policies, it is the treatment as well as the terms which matter, and the conditions of Government building contracts are, in one's experience, for the most part administered fairly and justly.

These conditions are primarily used by the Property Services Agency of the Department of the Environment although other contracting Government departments could also do so.

9.1 Form GC/Works/1 – contract documents

The documentation differs considerably from the Standard Form of Building Contract and before reviewing the terms of GC/Works/1 one must consider the contract documents and tendering procedure. All references in brackets (eg, 1.1) are to the conditions in GC/Works/1.

The contract documents are defined in Condition 1 to include those documents forming the tender and acceptance, GC/Works/1 conditions, the abstract of particulars and supplemental clauses, two special addenda affecting programming and in addition the specification, bills where provided, and the drawings of which three copies of each are to be made available to the contractor (1.1). In the case of a discrepancy, the conditions prevail over specification, bills or drawings.

9.2 Tendering procedure

Tracing through the tendering procedure sheds more light on the contract documents.

(a) Tenders are invited on Form C1009T on the terms set out.

(b) The tenderer states that he has perused all the relevant documents (eg, GC/Works/1, abstract of particulars, specification, bills (where provided), drawings, etc), accepts the selective tendering conditions and offers to carry out the work for the sum shown in the bills or the alternative basis provided. Acceptance makes a binding contract.

(c) Form GC/Works/1 sets out the conditions under which the work will be carried out but it does not contain articles of agreement or an appendix as in the Standard Form and is not signed by the parties. The abstract of particulars (C1009 Abs) which is part of the tendering documents is referred to in several of the GC/Works/1 conditions and provides the terms of the contract on completion, damages, maintenance period, etc. Alternative forms are used for firm price and for variation of price contracts based on the formula method (giving base month, etc – supplementary condition 146 to apply).

(d) There are also two addenda – one listing dates after acceptance for the supply of certain information relevant to condition 53.2.a and .b (prolongation and disruption) and the other giving the periods from acceptance by which the contractor requires subcontract nominations.

(e) There is also a supplementary condition No 139A on value added tax (C1953A – September 1975) which the authority agrees to reimburse to the contractor.

To establish fully the terms of the contract one may therefore have to consult up to ten different documents.

The basis for pricing may be bills (SMM based, provisional or approximate quantities) or schedules of rates. Current PSA policy is to provide bills for work estimated to cost more than £30,000, but for straightforward pre-planned projects up to £50,000 in value the basis may be specification and drawings only, the contractor to provide a schedule of rates for pricing variations.

Bills of quantities, unless otherwise stated, are deemed to be prepared on the method of measurement indicated (5.1). Any error in description of quantity or any omission will not vitiate the contract, and the error will be rectified, treated as a variation and the contract sum adjusted accordingly. Errors in the contractor's prices are not to be rectified (5.2). In provisional bills or bills of approximate quantities, quantities given will not limit the extent and nature of the work to be done (5.3). Where bills are not provided the references to bills in the conditions are cancelled. A schedule of rates may be supplied by the authority and work to be executed will not be limited by its descriptions (5A). Where the authority supplies neither bills nor a schedule the contractor may be required to supply a full and detailed schedule of the rates which were used to calculate the contract sum (5B). The value of the whole of the work, executed to the satisfaction of the SO, will be ascertained by measurement and valuation in the case of provisional bills, bills of approximate quantities or a schedule of rates (10).

Definitions are set out in condition 1.2. 'Accepted risks' are those of fire, storm, aircraft, radiation, riot, civil commotion, etc accepted by the employing authority. The 'Authority' (eg, the Property Services Agency), the SO (superintending officer) and the quantity surveyor are all designated in the abstract of particulars. All notices to be given under the contract must be in writing (1.6).

The contractor cannot assign or transfer the contract without the written consent of the authority (27).

9.3 Site conditions

The contractor must satisfy himself fully on site conditions, including access, and on the conditions under which work will have to be carried out (eg, availability of labour and materials) except for the information required to be given under the method of measurement on which the bills have been prepared (2.1). Any misunderstanding or misinterpretation will not give grounds for a claim, nor will there be release from contract obligations on the grounds that any matter was not, nor could have been, foreseen (2.2). (It will be noted that this condition is very widely drawn and places a very considerable responsibility on the contractor.)

9.4 Vesting of works

Condition 3 calls for special comment since it provides that throughout the whole period of the execution of the contract 'the works and any things brought on the site' (ie, plant, equipment, materials for incorporation) and owned by the contractor shall 'become the property of and vest in the Authority' (subject to a right of rejection), the contractor to remain responsible for protection and preservation (see also condition 26.1). The decision of the SO is final and binding on what unused 'things' the contractor may be permitted in writing to remove from the site (3.2). This rather unusual provision is designed to ensure that nothing which is required for the contract is removed from the site. It would appear that at the end of the contract the vesting provision would cease, plant, etc and surplus materials reverting to the contractor. This provision does raise complications in regard to prior title and reputed ownership which may defeat the vesting provision.

9.5 Carrying out the work

9.5.1 SETTING OUT

The SO is to supply detailed drawings, levels and other information necessary to enable the contractor to set out the works. The contractor is to set out accordingly, providing and maintaining the requisite instruments, equipment, etc (12).

9.5.2 FOUNDATIONS, ETC

Excavations must be examined and approved by the SO before any foundations are laid (21) and notice must be given before any work including variations is to be covered in with earth or otherwise (22).

9.5.3 MATERIALS, ETC

The contractor must, if requested, prove to the SO that the goods and materials are of the standard required in the contract documents (13.1). The SO has power to inspect workmanship and materials on site or at any factory or workshop, the necessary facilities to be given by the contractor (13.2). The contractor must provide facilities for the SO to carry out tests. An independent expert may be employed at the discretion of the SO, his report to be final and conclusive and the cost borne by the contractor if he is at fault (13.3). Defective work or materials must be rectified by the contractor (13.4).

9.5.4 STATUTORY NOTICES

The contractor must give, and meet the cost of, all statutory notices including fees (eg, local authorities, public utilities) (14). These must be included in the tender since there is no provision for recoupment.

9.5.5 DAMAGE TO PUBLIC ROADS

Provided the contractor takes all reasonable steps to prevent damage to highways, roads and bridges by extraordinary traffic by selection of routes, choice and use of vehicles and load distribution, and complies with any instructions of the SO, the authority will indemnify him for any claims for damage by extraordinary traffic provided the claim is promptly notified to the authority for settlement (48).

9.5.6 SITE CONDITIONS

The contractor is responsible for protection, security, lighting and watching and must provide the watchmen necessary (17). He must keep the site free from rubbish (19) and take all reasonable precautions to prevent a nuisance to neighbours or the public (18). Proper fire, etc precautions must be taken (25). All the material excavated remains the property of the authority to be dealt with as the contract conditions or the SO directs. Where fossils and antiquities are found the contractor must tell the SO at once and take every precaution to preserve them and not to disturb them and to cease work if this would endanger their excavation or removal (20). There is no provision for recompense in this condition but condition 9.2 may support a claim.

9.5.7 FROST

The SO may require work to be suspended to avoid damage by frost, inclement weather, etc. Only if the contractor has fully complied with the specification relating to frost, etc damage can he claim for the expense incurred (23).

9.5.8 DAMAGE

Plant and equipment, etc not for incorporation in the works are at the sole risk of the contractor who must immediately make good any loss or damage (26.1).

Damage to the works or to materials for incorporation must promptly be made good at the contractor's expense except where wholly caused by any of the accepted risks (fire, storm, aircraft, radiation, riot, etc) or by the neglect or default of a servant of the Crown when the authority will meet the cost or, if partially responsible, share the cost (26.2.).

9.5.9 COMPLETION: EXTENSIONS OF TIME

On or before the completion date the works must be completed, cleared of rubbish and delivered up to the satisfaction of the SO (28.1). Provided the contractor is not at fault, reasonable extensions of time should be allowed by the authority, but immediately he is aware of actual or possible delay the contractor must notify the SO of the circumstances and the actual or estimated extent of the delay. He must of course use his best endeavours to prevent or minimise delay. Extensions can be for one or more of the following reasons under condition 28.2:

(a) execution of any modified or additional work (time saved by omissions may be taken into account)

(b) weather conditions making continuance of work impracticable (ie, the con-

tract or a section of it has been completely stopped)
(c) act or default of the authority
(d) strikes or lockouts – including those engaged in preparing and manufacturing materials, etc (but not apparently in transport)
(e) any of the accepted risks (defined in condition 1.2 as fire, storm, etc, aircraft, radioactivity, riot, etc)
(f) any other circumstances which could not have been foreseen when the contract was entered into and over which the contractor has no control.

Under (f) there would be a basis for a request for an extension of time where a direct contractor of the authority held up the main contractor or where labour, for example, was short because of a competing major contract in the vicinity of which there was no knowledge at the time of tendering. Delays by nominated subcontractors or suppliers are unlikely to be considered under this heading.

9.5.10 PARTIAL POSSESSION

Part of the works certified by the SO as completed to his satisfaction may be taken over before the whole is completed if this part is specified in the abstract of particulars, if the parties agree, or the SO instructs. This part, the value of which must be certified by the SO, will be regarded as completed for the purposes of the provisions on vesting (see above), the maintenance period, release of retention money and damages (abated proportionately). The decision of the SO under this condition is final and conclusive (28A).

9.5.11 DAMAGES

Once the authority has given notice to the contractor that he is not entitled to an extension of time (or a further extension), damages are recoverable at the rate set out in the abstract of particulars for the period until the works are completed and the site cleared.

9.5.12 DEFECTS

Defects in the works due to failure or neglect on the part of the contractor or any subcontractor or supplier appearing in the maintenance period specified in the abstract of particulars (usually six months for general building works; 12 months for specialist engineering services) must be made good to the satisfaction of the SO at the contractor's expense (32.1). An exception is frost damage appearing after completion, provided the cause was after that date. If the contractor does not make good the defect the authority may do so and recover the costs from the contractor (32.3).

Where there is a separate subcontract maintenance period and defects have been made good within it, the contractor's responsibility will continue until the end of the period or six months from making good whichever is the later.

9.5.13 EMERGENCY POWERS

If the SO considers that urgent measures are called for because of, or to avoid, any risk or accident in connection with the Works, the authority may carry out this work if the contractor is unable or unwilling to do so, the cost to be recovered from him where he is liable (49).

9.5.14 OTHER WORKS

The authority is empowered to have other work done on the site while the contract is in progress, the contractor to give reasonable facilities for this purpose (50). The contractor, if he is not at fault, is safeguarded against any damage which results, or expense caused by delay to the contract works. Claims may be made under condition 28 for extensions of time and under condition 53 for prolongation and disruption expenses.

9.5.15 INSTRUCTIONS OF THE SUPERINTENDING OFFICER

The contractor is given notice for the possession of the site or to commence work (6) and he must carry out and complete the work to the satisfaction of the SO (7.1) (ie, the completion date is known at the outset but the starting date remains to be fixed). The SO may issue instructions in writing or orally (to be confirmed within 14 days if requested by the contractor (7.2)), and the decision of the SO is final and conclusive on whether they are necessary or expedient (7.3). If after a notice from the SO the contractor fails to comply with an instruction, the authority may have the work carried out, the additional cost to be recovered from the contractor (8). Condition 7 brings together the instructions which the SO may give including:

(a) further drawings, details and instructions varying or modifying the design quality and quantity of the work
(b) adding, omitting or substituting work
(c) correcting discrepancies (specifications/bills/drawings)
(d) removal or substitution of materials
(e) work – removal, order of execution, re-execution, suspension, opening up for inspection, emergency work
(f) making good defects
(g) working hours, overtime, replacement of foremen.

The net is cast widely by a final provision on 'any other matters . . . necessary or expedient'.

The contract sum is adjusted for the value of alterations, additions or omissions resulting from a written instruction of the SO. The addition or deduction is to be ascertained by the quantity surveyor on bill (or schedule) rates for similar work, or if not applicable, based on them where practicable, otherwise by measurement and valuation at fair rates. Failing this the daywork rates quoted in the contract will apply. Where valuation on bill rates (or based on them) would, in the opinion of the quantity surveyor, not give a reasonable figure, measurement and valuation at fair rates are to be used (9.1).

The SO must be given reasonable notice of the commencement of any daywork ordered and within one week of the end of each pay week the vouchers for labour, materials and plant must be delivered to the SO in the form required (24).

If in complying with a written instruction other than on alterations, etc the contractor, through no default on his part, incurs direct expense not reasonably contemplated by or provided for in the contract, the amount as ascertained by the quantity surveyor is to be added to the contract sum and included in interim payments without the deduction of retention money. As a condition precedent to this the instruction must be in writing, and the contractor must provide as soon as reasonably practicable the requisite documentation on the expense incurred. By the same token any saving will result in a reduction (9.2).

The authority may appoint a resident engineer or clerk or works who will exercise the powers of the SO in respect of materials and workmanship under Condition 13.1 and 13.2 and such other powers as the SO notifies to the contractor (16).

9.6 Subcontracting

9.6.1 SUBLETTING

The previous consent of the SO is required for any subletting (30). In addition to incorporating the power to determine for the same reasons as those contained in Condition 44, the main contractor must include in the subcontract and ensure the observance of provisions as follows:

(a) all materials, etc for incorporation to vest in the contractor
(b) the contractor must be able to fulfil his obligations to the authority on removal of materials, etc from site (3.2), return of drawings, etc (4.4), inspection and tests on conforming to description (13.2 and 3) on regulations (35), fair wages (51), admission to sites (56) passes (57) and photos (58)
(c) impose conditions similar to those placed on the contractor on assignment (27), replacement of employees (36), corruption (55) and secrecy (59)
(d) consent of contractor to be essential to subletting
(e) price variation for labour-tax changes (11G).

9.6.2 NOMINATION

PC sums apply to subcontractors and suppliers nominated by the authority, the contractor before placing the order having the right of reasonable objection (31.1). Whether nominated or approved by the authority or the SO, or appointed under their direction by the contractor, full responsibility falls on the contractor who must make good any loss or expense suffered by the authority by reason of default or failure (31). This overall responsibility for the nominated subcontractor and supplier seems to extend beyond such circumstances as delays, liquidation, etc. It could even include design faults, etc if part of the subcontractor's responsibility, and for this reason a nomination on this basis should be considered unacceptable (31.1). The consent of the authority is needed for the contractor's selection of another subcontractor or supplier to complete the work or to do so himself. The PC adjustment will be on the basis of the amount of the *original* subcontract (38.5).

Cash discount is 2½ per cent on the total due, the actual terms to be agreed between the parties. Payment to the contractor for fixing, including unloading, etc, is to be at rates in the bills (or the schedule). Contractor's profit as priced is to be pro-rata to the prime cost adjustment excluding labour-tax variations (38.2). The authority may order and pay direct for any items covered by the PC sums, the contractor to be compensated for loss of profit, as priced, on the amount paid direct but not for loss of cash discount (38.4).

The onerous nature of these provisions makes it essential for the contractor to have nominated a subcontractor who is likely to meet fully his commitments. The 'authority' (ie, the PSA) helps to the extent that it will either invite the contractor to approve a list before tenders are invited for nomination purposes and amend it if specially desired, or ask the contractor on the details supplied to invite tenders from firms *he* regards as suitable to be sent direct to the PSA for nomination of a firm considered suitable on the basis of tenders received. Above all, the terms, including the programme, must be fully agreed at the outset. Detailed conditions for nomination are set out in the authority's tendering documents but there is no Government subcontract form for use in conjunction with GC/Works/1. However, in 1980 the NFBTE are issuing a suitable subcontract form for both nominated and non-nominated subcontract work and its provisions are reviewed in section 8.8.

9.7 Payment

9.7.1 ADVANCES ON ACCOUNT

Condition 40 requires the contractor at intervals of not less than one month to submit claims for payment on account of work done and materials reasonably brought on to site for incorporation and adequately stored and protected (but not off-site materials). Payment is 97 per cent of the value certified by the SO. Claims must be supported by valuations based on rates in bills, in the schedule of rates or an appropriate alternative basis. Where the contract sum exceeds £100,000 the contractor may apply for fortnightly interim advances on account.

The SO decides on the amounts to be included on account of entitlements under conditions 9.2 (extra direct expense incurred in complying with written instructions of the SO) or 53.1 (a new provision for prolongation or disruption expenses – see below). These are not subject to retention.

No interim certificate is conclusive and it may be subsequently modified or corrected (42). The authority's decision is final and conclusive on the right to an interim certificate and sums certified.

If, before payment of any advance or the issue of the final certificate, the contractor is unable to satisfy the SO on request that amounts due to subcontractors or suppliers covered by previous advances have not been paid, payment of the amount in question may be withheld by the authority until the SO is satisfied (40.6). This could happen if, for example, the contractor frequently or substantially delayed payments. If in the case of a subcontractor or supplier nominated under a PC sum the

SO certifies that certain amounts have not been paid, the authority may pay direct and recover from the contractor. The decision of the SO or the authority is final and conclusive on these matters.

9.7.2 FINAL PAYMENT

The SO certifies amounts due as advances and as final payment. Once the SO has certified the date on which the works are completed to his satisfaction the contractor is entitled to be paid the final sum less half the reserve (ie, retention money of 3 per cent). As soon as possible after the completion certificate the quantity surveyor must send a copy of the final account to the contractor, who must provide all necessary documents and information and arrange if required for his representative to attend the site with the quantity surveyor for measurement purposes (37).

At the end of the maintenance period detailed in the Abstract of Particulars (usually six months) the SO, if satisfied, issues a certificate that the works are in a satisfactory state. The balance of retention (1½ per cent) is then due together with any balance in respect of the final sum (41). These provisions envisage the possibility of agreement and payment of the final sum before the end of the maintenance period but in practice the certificate of the SO at the end of the maintenance period would be awaited.

9.7.3 PROLONGATION AND DISRUPTION EXPENSES

A new condition 53 in the 2nd edition of GC/Works/1 makes provision for reimbursement of any expense properly and directly incurred by the contractor beyond that provided for, or reasonably contemplated by, the contract as a result of the regular progress of the works being materially disrupted or prolonged for the following reasons:

- (i) complying with any written instruction of the SO provided the contractor is not in default (cf, condition 9.2)
- (ii) making good loss or damage to the works caused by a Crown servant or through the execution of other work by the authority during the currency of the contract
- (iii) delay in providing drawings, etc: delay caused by work carried out direct by the authority; delay in making a nomination for which the contractor has given reasonable notice to the SO

The conditions precedent to such a payment include immediate notice by the contractor to the SO of the likely claim and of the circumstances giving rise to it. Documentation must follow as soon as reasonably practicable. The cost of making good any loss or damage to work for which the contractor is responsible is excluded (26.2)

9.7.4 FLUCTUATIONS

The abstract of particulars sets out the provisions for variation of price for all contracts ie, whether firm or variable price. Current Government policy is that contracts of less than 12 months' duration are let on a firm price basis. Condition 11G, which applies to all contracts, provides for adjustments to the contract sum for increases or decreases which come into effect after the date of tender due to labour tax matters (ie, tax, levy or contribution payable by law in respect of workpeople) insofar as they relate to *workpeople* – ie, skilled or unskilled, manual labour employed directly on site, including such labour as is chargeable to overheads. Where any workpeople are partly employed on site the adjustment is pro-rata. The contractor must ensure that the terms of his subcontracts provide for similar adjustments.

Where tenders are invited on a variation of price basis, Supplementary Condition 146 is used, price adjustment to be on the formula basis. This is fully explained in Chapter 4, the Government generally using workgroups instead of the more numerous work categories. (See Form 1960M for the allocation of work categories (Series 1, 34; and Series 2, 48) to the 13 work groups. (details in Appendix G))

Conditions 11A to 11F, which provide for the calculation of proven fluctuations, are now seldom used. Supplementary Condition 145, which sets out the provisions of 11A to F, covers recovery of the costs of labour (11A), materials (11B), subcontracts, other than nominated (11C), nominated subcontracts (11D), special arrangements (11E) and general provisions – eg, notice to SO (11F). The recovery provisions in 11A and 11B are similar in principle to those of clause 39 of the Standard Form (reviewed in Chapter 4) with the significant difference that in Condition 11A, no provision is to be made for increases in wages, etc coming into effect after the date of tender even if known at that date (clause 39 requires these to be taken into account).

Note: Supplementary Conditions 185, 186 and 187 (Counter-Inflation – Incomes Policy) introduced in April 1978 were withdrawn in December 1978.

9.7.5 RECOVERY FROM CONTRACTOR

Amounts recoverable from the contractor under the contract may be deducted from any sums due or which become due under this or any other contract with the Government (43).

9.8 Special provisions

There are several conditions, some special to government contracts, which can be summarised as follows.

9.8.1 INJURY TO PERSONS AND PROPERTY

A noteworthy feature of GC/Works/1 is that the contractor is not contractually required to provide insurance cover for employer's and public liability risks, but legislative and commercial requirements would ensure that this was done.

Condition 47 makes the contractor responsible for third party claims by or against Crown servants resulting from personal injury or loss of property other than the works, but arising out of their execution, and in particular for:

(i) reinstating or making good any loss of property suffered by the Crown
(ii) indemnifying the Crown for any claims made against it or its servants for personal injury or loss of property, and for any payment by the Crown by way of indemnity to a servant of the Crown or in respect of any loss of property or personal injury suffered by a servant of the Crown.

If the contractor can show that the personal injury or loss of property was not due to his neglect he will not be liable. The liability will be apportioned if he is only partially responsible. Indemnities must not exceed the amount which would be recoverable at law.

The authority must notify the contractor of any claims or proceedings arising and if he admits his liability to indemnify, he or his insurers must deal with the matter. Before any action is taken they must consult the legal adviser to the authority if the issue involves any privilege or special right of the Crown.

9.8.2 SUPERVISION

The contractor must employ a competent, full-time agent to receive directions by the SO and to supervise the execution of the works (33). Each day he must provide the SO with the number and description of men employed on the works (34).

9.8.3 EMPLOYEES

Where the SO considers undesirable the continued employment of any person from agent downwards, he may require the contractor to remove that person from the site and replace him (36). The decision of the authority or SO on this matter is final and conclusive.

9.8.4 EMPLOYMENT CONDITIONS

There is the usual fair wages clause (51) and a requirement that there will be no racial discrimination by the contractor, the subcontractors or their employees, etc (52).

9.8.5 SITE REQUIREMENTS

There must be compliance with the regulations of any Government establishment on which work is carried out (35). The authority may give notice to the contractor of any person not to be admitted to the site or to be admitted only with written permission. The contractor must also, if required by the SO furnish a list of names, addresses and designations of all persons concerned with the work. Any decision of the authority under this heading is final and conclusive (56). Where passes are required, the SO will arrange for their issue (57). The written permission of the authority is first necessary before any photographs are taken of the site (58). The attention of the contractor and of his employees is drawn to the Official Secrets Acts and the need to obtain the authority's consent to disclose to outsiders any information about the contract (59).

9.8.6 CORRUPT GIFTS

The contractor is forbidden to offer any gift, etc to a servant of the Crown as an inducement or reward in respect of any government contract or to pay commission unless full particulars of this were disclosed to the authority in writing before the contract was entered into. Where there is a breach of this provision the authority is entitled to determine the contract and/or recover the value of the gift from the contractor. The authority's decision in these matters is final and conclusive (55).

9.9 Determination

In addition to the power to determine because of corrupt gifts or payments or commission under condition 55, the authority may determine the contract under condition 45 due to the default or failure of the contractor and, also under condition 44 which contains special powers.

Note: There is no condition enabling the contractor to determine.

9.9.1 CONTRACTOR'S DEFAULT

The authority may by notice determine the contract in the following circumstances:

(a) failure to comply within seven days with a notice from the SO to remedy work which is defective or is being carried out inefficiently
(b) delaying or suspending work which, in the judgement of the SO will make it impossible to complete in time
(c) financial failure – bankruptcy or insolvency
(d) failure to comply with a site admission requirement prejudicial to the State

Determination is without prejudice to the authority's rights against the contractor for delay, defective work, breach of contract, etc (45).

9.9.2 SPECIAL POWERS

Without any reason being given, the authority may give notice on an entirely discretionary basis to determine the contract, and as soon as practicable (and not more than three months from the date of the notice or the period up to completion date, whichever is the shorter) give directions for prompt compliance by the contractor on such matters as:

(a) protecting the work or carrying out further work
(b) removal from site of plant, materials, etc or any rubbish
(c) termination or transfer of subcontracts and other contracts for plant hire, etc
(d) any other matters arising out of the contract.

In the event of determination the contractor is to be paid for work executed, materials available, incomplete subcontract work, plant hire, loss or damage to the works and any labour costs (eg, redundancy) resulting from the determination.

There is also provision for allowance for hardship suffered (including loss of profit) on which matter the authority's decision is final and conclusive (44). This clause reflects the exigencies including political situations affecting Government contracting without any reflection on the contractor's performance.

9.9.3 DETERMINATION AND SETTLEMENT

When a contract is determined because of corruption (55) or the default of the contractor (45) sums due or accruing due are 'frozen'. To complete the work the authority may either employ another contractor or use existing facilities, eg, hire the contractor's labour, take possession of the site and all 'things' on it (materials, plant, etc) or purchase or do anything needed for completion, without the right to a claim by the contractor. The contractor, except where bankrupt or in liquidation, must, if required by the authority, assign without payment the benefit of subcontracts for work and materials, the authority to assume financial responsibility. The authority may pay direct to nominated subcontractors and suppliers unpaid amounts already certified and recover from the contractor.

The cost of completion is certified by the SO including any liquidated damages due from the contractor at the date of determination. If this cost, plus payments to the contractor, is less than what the contractor would have been paid for due completion, he receives the difference subject to the computation in condition 46.2. Where there is an excess the authority may first sell 'things' on the site (to which the contractor has a title) the balance then owing, to be recovered from the contractor (or if there is a residue after the sale the balance is to be paid to the contractor) (46.3).

9.10 Arbitration

All disputes or differences may be referred to a single arbitrator agreed between the parties, failing which the authority may request an appointment by the President of the Law Society, RIBA, RICS, ICE or one of the specialist engineering institutions. Unless there is an agreement to the contrary arbitration must await completion, abandonment or determination. There are certain important exclusions from this right to arbitrate.

(i) matters relating to fair wages – condition 51

(ii) any matter where the decision of the authority or the SO is expressed to be final and conclusive – eg, removal of 'things' from site (3.2); necessity for SO's instructions (7), quantity surveyor's valuations under 9.3; price variation (11G); independent report on materials (13); SO's decision on partial possession (28A) or on replacement of contractor's employees (36); on direct payment to nominated subcontractors or suppliers (40); on right to or amounts certified in interim certificates (42); hardship on special determination (44); certain costs on determination under 46.1.e; corrupt gifts (55); admission to the site (56).

9.11 Comparison with Standard Form

In considering the provisions of GC/Works/1 there have been several references to the Standard Form of Building Contract which have highlighted some significant differences.

These points are brought together in this brief summary which covers fundamental distinctions and several important but less vital points.

9.11.1 FUNDAMENTAL DISTINCTIONS

There are three main areas of difference from the Standard Form – nomination, determination and arbitration.

Once the nomination of a subcontractor (or supplier) has taken place and been accepted by the contractor he is fully responsible for the carrying out of the work covered by the nomination. The PC adjustment will be on the basis of the original contract sum; any costs arising from delay, bad workmanship or failure to carry out the subcontract, including the bankruptcy or liquidation of the subcontractor, must be borne by the contractor.

There is no provision for the contractor to determine the contract and the authority, in addition to the usual powers to determine, may also determine for special reasons unrelated to the contract performance of the contractor.

The arbitration clause is reduced in scope by excluding from references to an arbitrator ten or more matters on which the decision of the authority and the SO are 'final and conclusive'.

9.11.2 IMPORTANT DIFFERENCES

The differences in documentation have been dealt with at the beginning of this section.

In the execution of the work, GC/Works/1 requires examination and approval of the SO, before any foundations are laid and notice must be given before work is covered in with earth. The SO may require work to be suspended to avoid damage by frost or inclement weather, with the contractor to be recompensed if he is not in default. Materials and contractor's plant on site are vested in the authority whose written permission is required for removal. Completion must be entire (not 'practical') ie, the works delivered up to the satisfaction of the SO with all materials, plant and rubbish cleared from the site. The instructions of the SO cover a wide field and the necessity for them cannot be contested. They are not required to be in writing but the contractor can ask for written confirmation of any instruction within 14 days – and would be wise to do so. The authority may arrange for other work to be carried out on the site at the same time and may order work covered by a PC sum to be carried out direct, the contractor to be paid for loss of profit.

The reasons for extensions of time do not specifically include delays because of shortage of labour or materials or resulting from the work of a direct contractor of the authority, although there would be a basis for a claim under 'unforeseen circumstances outside the contractor's control'. Only if work is actually stopped by bad weather is an extension possible for this reason. There are generally no extensions of time for delays attributable to nominated subcontractors or suppliers.

In the matter of payment, retention is only 3 per cent, with provision for fortnightly payments in contracts over £100,000 in value. Although amounts agreed for extra expense on valuation of variations (9.2) and prolongation or disruption (53.1) are to be included, without deduction of retention, in monthly advances, off-site materials are not allowed for. The cash discount of 2½ per cent applies to both nominated subcontractors and suppliers. There is no provision for statutory fees, etc but the cost of preservation of antiquities is to be added to the contract sum. Over-payments will be adjusted and amounts due from contractors deducted from payments under *any* contract with the Government. Payments may be made direct to nominated subcontractors and suppliers where sums due have been withheld by the contractor. The authority may also hold back payments where satisfied that subcontractors or suppliers (not necessarily nominated) have not been paid.

There are special provisions mainly peculiar to Government contracting. There is no contractual requirement to insure for employer's or public liability. There are stipulations on employees (suitability etc), admission to sites, passes, photographs; the provisions of the Official Secrets Acts, and requirements on emergency action, conforming to regulations, corrupt gifts, etc.

9.12 Form GC/Works/2

There are 19 conditions in this relatively simple form.

A firm price is submitted by the contractor on the basis of an abstract of particulars and, usually accompanied by a specification and drawings. The SO may order in writing alterations, additions and omissions, and the valuation is to be on the basis shown in the tender, that is, on rates and prices in a tendered schedule of rates or in a schedule of rates prepared by the contractor on the basis of his tender prices.When values cannot be ascertained from the schedule of rates or deduced from it, daywork at the rates set out will be used.

Payment intervals of not less than one month will be at the contractor's request. The payments comprise: 97 per cent of value of work and of the cost of materials reasonably brought onto the site as certified by the SO. Half the balance is payable on completion, the other half at the end of the maintenance period set out in the abstract of particulars which also stipulates the liquidated damages in the event of any delay after the given completion date. In contracts over £10,000 in value a variation of price for labour tax matters may be allowed under supplementary condition 119A.

Other clauses deal with indemnity to the Crown for injury to persons or loss of property, determination of contract, fair wages, subletting, admission to site, secrecy and arbitration.

9.13 Measured Term Contracts

The continuing maintenance and other requirements of Government installations and establishments involving the services of builders calls for special contract provisions – 'General Conditions of Contract for Measured Term Contracts' Form C1501 (January 1974) with the accompanying documents which are reviewed under this heading.

As will be evident from the information given below, this work calls for special techniques and management requirements. Contracts are generally let through the Property Services Agency. Since one can regard the basis of 'pricing on or off' a schedule of rates as the outstanding feature of term maintenance contracts, this is dealt with first, a brief description then following of provisions for payment and other contract conditions.

9.13.1 PRICING

In the invitation to tender (C1521 A(T) September 1976) information is given of the estimated approximate annual value of the work in the contract area. It is made clear that unless the contractor agrees, no single work for which an order is placed will normally take more than 12 months to complete. The basis for tendering is set out in the invitation to tender as follows:

(a) The basis is a Schedule of Rates for Building Work (1973) which covers a wide variety of work divided into main trade sections. Each month in Addendum A it is updated by a single weighted percentage calculated from current labour and materials prices. This is designed to give effect to monthly price increases which have occurred since the schedule of rates was prepared but it will not fully reflect local price levels which the tenderer must take into account. The up-to-date percentage for any month will apply to orders placed during that month irrespective of the period for execution (ie, price fluctuations during the currency of the order will not be covered).

(b) On this basis the contractor submits a tender for work which will consist of a series of orders from the SO, within a minimum and maximum value of any single order from which the authority will not normally depart. The minimum period of the contract is generally six months, and the maximum three years, from the date of the letter of acceptance. During the period of notice the contract may be terminated at the end of six months or at any time thereafter on eight weeks' notice; the value of orders which can then be given is limited to half that of the preceding 13 weeks (condition 3).

(c) Pricing is by means of percentage adjustments on the current updated schedule of rates (see (a) above). Adjustments are in three categories.
   (i) *Primary percentage adjustment*: If the contractor thinks the official rates do not fully reflect his costs including overheads and profits, fluctuations during the currency of an order, or differences between Addendum A updating and local prices (but excluding VAT), he will calculate his percentage (on or off) accordingly.
   (ii) *Secondary percentage adjustment*: In orders (other than in the £500 to £2,500 measured work band, for which this adjustment is nil) the contractor may quote a further adjustment reflecting supervision costs, productivity levels and such factors as no payment on account for

work not exceeding £500. These percentages are to be in respect of each of the following bands: up to £100; £100/£500; £2,500/£10,000; over £10,000.

(iii) *Self measurement percentage adjustment*: On orders above a quoted value the contractor may, if the authority agrees, measure the work himself and on the SO's verification be paid accordingly (supplementary Condition 50.A). If the contractor makes excessive errors in his favour there would be a reversion to the normal arrangements for payment (see below). With this in operation the contractor may quote an additional percentage adjustment on the secondary value (as calculated in (b) above).

There is also provision for supplementary lump sum offers for minor repairs and maintenance of married quarters for six month periods, payable monthly.

9.13.2 PAYMENT

The contractor is entitled to claim, in the form required, payments at intervals, not less than monthly, for the full value of work (excluding unfixed materials) which is executed to the satisfaction of the SO on any order exceeding £500 (14). On completion of each order an account in the required form must be presented by the contractor within 28 days of final measurement (15).

Valuation of the work is by measurement and valuation in accordance with the basis of pricing outlined above and as set out in Supplementary Condition No. 51 Part 1 C1509 October 1975.

9.13.3 OTHER CONDITIONS

The successful contractor carries out work on the basis of the written orders of the SO which state a time for completion under condition 4. There is provision for extension of time where the contractor is not at fault. The completion date must be certified by the SO and variations, etc must be in writing. If work is ordered to which the rates do not apply, the valuation is based on rates and prices 'deduced therefrom'. If neither basis is suitable it becomes a matter for agreement between the contractor and the SO, failing which contract daywork rates apply (11). Notice of commencement of work to be executed as daywork must be given to the SO and weekly prime cost details forwarded by the end of the following week (13). Percentage additions to prime cost (RICS/NFBTE Definition 1 December 1977 as amended by PSA C1522 October 1976) and RICS Basic Plant Charges July 1975 (see section 8.7) are those set out in the tender by the contractor.

The authority reserves the right to use another contractor or its own personnel for work in the contract area (eg, armed forces) and to supply all or any of the materials required (2).

Subletting, with consent, is allowed under condition 10.1 and the contractor may be required to place work with a nominated subcontractor at 2½ per cent cash discount with 5 per cent to cover profit, facilities, etc. 'Particular Clauses to the Contract' (C1523 October 1975) deal with operational requirements – use of site, noise control, site facilities, scaffolding and plant, electrical supplies and installations, safety, security, fire precautions, etc. There is also the stipulation that all mechanical and engineering services are to be the subject of separate contracts with the authority and clause 7 sets out the procedures for other specialist works and supplies.

There are clauses similar in intent to those in GC/Works/1 covering a wide variety of points including materials and workmanship (22); indemnities to the Crown for injury to persons and the loss of property (40); damage to plant, works, building, roads, etc (37 and 39); determination for bankruptcy, etc (46) or default (47); fair wages (43); racial discrimination (44); secrecy (32); admission to site (29); passes (30); and arbitration (48).

APPENDIX A

ORGANISATIONS PARTICIPATING IN BUILDING CONTRACT MATTERS

The following organisations participate in the work of the Joint Contracts Tribunal and the National Joint Consultative Committee for Building. The years in which they joined the JCT and NJCC are given in brackets.

The Royal Institute of British Architects (RIBA)

Founded in 1834 and granted a Royal Charter in 1837, the RIBA has some 27,000 corporate members qualified by examination. Approximately half of them are in private practice. It 'exists to promote the highest standards in architecture and to conserve or improve the architectural environment'. The elected Council works through four Boards and the Institute has a code of professional conduct. (JCT 1931; NJCC 1955)

The Royal Institution of Chartered Surveyors (RICS)

The RICS, founded as the Institute of Surveyors in 1868, has over 39,000 members qualified by examination. There are six main divisions. The quantity surveyors division consisting of 11,600 members is concerned with building industry matters and contract conditions as well as the SMM. (JCT 1952; NJCC 1955).

National Federation of Building Trades Employers (NFBTE)

Representing the very substantial majority of firms engaged in contracting, the NFBTE was formed in 1878. It has some 11,000 members and its interests and functions are extensive and include wage negotiations. There is a National Contractors Group with 100 members and a 2,700 strong Federation of Building Subcontractors which, together with the main contracting interests of general contractors throughout the country, is concerned in contract conditions. (JCT 1931; NJCC 1955)

The Federation of Associations of Specialists and Subcontractors (FASS)

The Federation includes some 20 specialist associations involving trades such as flooring, roofing, plumbing, painting, tiling, shopfitting, piling, asphalt and concrete work. (JCT 1967)

The Committee of Associations of Specialist Engineering Contractors (CASEC)

This body, founded in 1961, represents constructional steel, heating and ventilating, and electrical contractors. (JCT 1966; NJCC 1977)

Association of Consulting Engineers (ACE)

Founded in 1903, this association has over 900 members in private practice as civil, mechanical, electrical, structural, etc, engineers. (NJCC 1973; JCT 1974)

Local Authority Associations

These consist of the Greater London Council, the Association of County Councils, the Association of Metropolitan Authorities, and the Association of District Councils. (JCT 1957)

The Institute of Building (IOB)

This institute was founded in 1843 and now has a membership of over 25,000. It was associated with the issue of the first Standard Form of Contract in 1903 but later withdrew. It is now primarily a professional examining institution in building practice.

The Building Economic Development Committee

This is a Government financed body which has had a direct influence on contract matters. This organisation produced the price adjustment formula discussed in Chapter 4. One of 17 EDCs set up by the NEDC, its 25 members are drawn from professional, industrial and trade union interests together with representatives of the Treasury and the DOE. It is not represented as such on the JCT or NJCC.

APPENDIX B

THE JOINT CONTRACTS TRIBUNAL

When a new edition of the Standard Form of Contract was issued in 1931, the JCT was constituted to keep the form up-to-date and to amend it as necessary. Over the years the JCT has provided this invaluable service and revised editions of the form were published in 1939, 1963 and now in 1980. Frequent reprints incorporate amendments which may be published in July each year, unless an urgent amendment is needed.

Initially, the JCT consisted of representatives of RIBA and NFBTE, who were followed by the RICS in 1952 and the Local Authority Associations and the LCC (now GLC) in 1957. In 1966 and 1967 subcontracting interests were represented when CASEC and FASS joined. In 1966 the Scottish Building Contract Committee (see section 3.1) joined the JCT and in 1974 ACE. (Association of Consulting Engineers). The CBI has 'observer status'. The Joint Secretaries are lawyers on the staff of the RIBA and NFBTE.

Practice Notes are issued to deal with problems arising from the use of the form. These are expressions of opinion on the intent and meaning of the contract and are generally accepted but only an arbitrator or the courts can give authoritative rulings. The most recent series of Practice Notes relates to the 1963 edition. These are summarised in Appendix C. The earlier series, containing 31 Practice Notes dating from December 1946 relate to the 1939 edition. A special Explanatory Memorandum has been issued on the 1980 forms.

Originally the JCT was only concerned with main contract conditions but with the inclusion of the subcontractors' organisations a Standard Form of Nominated Subcontract, previously published by the contractors' organisations (NFBTE,etc) is now issued by the JCT (see Chapter 6). This co-ordination under one authoritative body will increase the status and use of these forms and others issued under its authority, including the 1980 Design and Build Form (see section 7.5). The Formula Rules are now published in one booklet by the JCT (see section 4.4.1).

The Tribunal advises and consults with other bodies on contract practice and provisions. The initiative for amendments can arise either within the Tribunal or at the request of any constituent body, through which other bodies should raise any matter. Decisions are by common consent and this may call for a reconciliation of contrasting points of view.

The JCT issues an Annual Report reviewing its work. In the 1977 report the Chairman said that liaison was being maintained between the Tribunal and the DOE on EEC proposals affecting building and construction contracts. Comments from private sector property owners and developers have been available through an observer from the British Property Federation nominated by the CBI.

APPENDIX C

JOINT CONTRACTS TRIBUNAL PRACTICE NOTES

The 1963 edition of the Standard Form of Building Contract with subsequent amendments, has been the subject of Practice Notes issued from time to time by the Joint Contracts Tribunal. These deal with problems arising from the use of the Standard Form but are only expressions of opinion on intent and meaning, the arbitrator or the courts alone giving authoritative rulings.

The notes are extensive and in listing them below a brief indication is given of the subject matter. These have some relevance to the 1980 edition where the contract provisions are unchanged. An Explanatory Memorandum has been issued by the JCT on the 1980 forms and doubtless in due course a new series of Practice Notes will evolve.

Practice note 1

CLAUSE 19(2)(a) – INSURANCES
Requirement on contractor to insure for damage to adjoining property caused otherwise than by the contractor's negligence – explanation of provisions of clause – provisional sum in bills to cover insurance premium. (Revised July 1971).

Practice note 2

CLAUSE 23(j) – EXTENSIONS OF TIME – AVAILABILITY OF LABOUR, GOODS AND MATERIALS
Deletion of this optional clause must be very carefully considered at tendering stage having regard to risks falling on contractor (Revised March 1975).

Practice note 3

CLAUSE 31 – SURCHARGE ON IMPORTED MATERIALS
Import surcharge to be treated in same way as any other increase or decrease in the price of materials, etc (January 1965) (clause amended 1967).

Practice note 4

CLAUSE 31 – FLUCTUATIONS IN RATES OF WAGES OF BUILDING OPERATIVES.
Wage rates 'applicable to the works and current at the date of tender' relate to those which are the subject of a decision of the wage fixing body at the date of tender (ie, 10 days before the date for receipt of tenders) – (July 1965).

Practice note 5

CLAUSE 31 – TIMBER USED IN FORM WORK
Timber used in formwork should be the subject of a price list adjustment where the timber is shown on the basic price list (July 1965) (clause amended 1967).

Practice note 6

CLAUSE 28 – NOMINATED SUPPLIER: TERMS OF CONTRACT – 5 PER CENT CASH DISCOUNT
The instructions of the architect should be sought where the contractor is required to place an order without the 5 per cent cash discount if this entitlement is to be preserved (July 1965) (clause amended 1972).

Practice note 7

CLAUSES 3 AND 5 – DRAWINGS – ADDITIONAL COPIES
Additional copies as required should be provided at cost (July 1965)

Practice note 8

CLAUSE 30 – CERTIFICATES AFTER ISSUE OF CERTIFICATE OF PRACTICAL COMPLETION
Further certificates before the final certificate may be issued if justified (July 1965) (clause amended 1972).

Practice note 9

Withdrawn.

Practice note 10

CLAUSES 30(2)(A) AND 14(2) – PAYMENT FOR OFF-SITE MATERIALS AND GOODS
Although it is a matter for the architect's discretion the discretionary power to include the value of off-site goods in certificates should be exercised particularly where goods are being specifically fabricated. (Revised January 1975).

Practice note 11

CLAUSE 31 – REDUNDANCY PAYMENTS ACT 1965
Contributions through the National Insurance Scheme come within the scope of this clause but the clause is amended to clarify the point. (April 1966).

Practice note 12

LONDON GOVERNMENT ACT 1939
Minor amendment relating to Local Government legislation. (April 1966).

Practice note 13

APPLICATION IN SCOTLAND OF PRACTICE NOTES 1-12
Following adoption of the Standard Form in Scotland in October 1965, certain consequential amendments to Practice Notes are set out in the light of Scots Law. (April 1967).

Practice note 14

CHANGEOVER TO METRIC – CLAUSE 11
Procedural points on changeover to metric – the period 1 January 1969 to December 1973 when no new work should be in imperial terms. (December 1968).

Practice note 15

CLAUSE 31 – FLUCTUATIONS – USE OF A OR B – FIRM PRICE – LIST OF MATERIALS
Emphasises that either Clause 31A or 31B should be deleted – draws attention to the limited fluctuations on listed materials under B, ie, rise or fall due to duties or taxes. (April 1969).

Practice note 16

FIXED FEE FORM OF PRIME COST CONTRACT

Disputes on 'nature or scope of the works' can be reduced by a comprehensive specification, the use of a provisional sum if necessary, a realistic estimate of prime cost. (May 1969).

Practice note 17

VALUE ADDED TAX

Sets out fully VAT provisions in relation to building work and the treatment of VAT under the Standard Form. Appendix A sets out Amendment 7/1973 and the VAT Supplemental Agreement. Appendix B gives notes on VAT Clause 13A. Appendix C gives notes on and a specimen of an authenticated receipt. (February 1973).

Practice note 18

ADJUSTMENT OF THE CONTRACT SUM BY MEANS OF THE FORMULAE

Explanation of the operation of the formula provided for in clause 31F: formula rules, nominated subcontracts, tender price level, productivity deduction, interim valuations, etc. Appendix A gives notes on clause 31F. Appendices B, C, D and E relate to specialist lift, structural steel, electrical, heating and ventilating and air conditioning installations (March 1975).

Practice note 19

CLAUSES 19, 19A, 20 AND 30(2A) – INSURANCE PROVISIONS

Deals with insurance for 'full value' under clause 20(a)(1), public and employers' liability amendments to clause 19(1)(a), clause 20 and a new clause 19A excluding risks for nuclear perils (separately dealt with by statute), clause 20(A) insurance cover on determination of employment of contractor, and clause 30(2A) insurance of off-site goods and materials (July 1975).

Practice note 20

STANDARD FORM OF BUILDING CONTRACT FOR USE WITH BILLS OF APPROXIMATE QUANTITIES

Explains the main departures from the Standard Form and in the Appendix lists the differences under each clause (October 1975).

Practice note 21

SECTIONAL COMPLETION SUPPLEMENT
For use only where tenderers are notified that work will be carried out and taken over by the employer in phased sections. Sections must be clearly identified. Modifications of contract reviewed. Need to reach agreement with insurers on extent of cover under clause 19(2)(a) for sections completed and handed over. (December 1975).

Practice note 22

FINANCE (NO 2) ACT 1975 – TAX DEDUCTION SCHEME: NEW CLAUSE 30B
Describes the scheme operating from 6 April 1977 and its effect on the Standard Form. The amendment (clause 30B) is set out. It makes provision for the situation where the employer is a 'contractor' and the contractor a 'subcontractor' under the scheme. (November 1976).

Practice note 23

ADJUSTMENT OF THE CONTRACT SUM BY MEANS OF FORMULAE (SERIES 2)
This Practice Note includes material published in Practice Note 18 adapted for application to Series 2 (48 work categories) dated 4 April 1977. It explains the Series 2 rules and deals with nominated subcontractors, extent of application of formula, tender price level, interim valuations, etc. Appendix A gives notes on clause 31F (amended April 1977) and Appendices B, C, D and E deal with certain aspects of the special indices. (November 1977).

APPENDIX D

STANDARD FORM OF BUILDING CONTRACT CLAUSE HEADINGS – COMPARISON OF 1963 AND 1980 EDITIONS

In the 1980 edition of the Standard Form the Joint Contracts Tribunal has substantially changed the format and introduced several new clauses, decimal numbering now being used throughout. Below are listed the main clause headings to give the broad relationship between clauses in the 1963 and 1980 editions and to provide a readily available picture of the contents of the new edition. The numerous subclauses (over 150, including 26 in clause 35 alone) have not been included but these are listed in the contents section at the beginning of the 1980 edition itself. The conditions are prefaced by the Articles of Agreement but in the 1980 edition a new Article 5 sets out the provisions for arbitration (clause 35 – 1963 edition). The details of the 1980 edition are fully considered in Chapters 3 (general), 4 (fluctuations) and 6 (nominated subcontractors).

Clause number 1980 edition	Clause heading	Clause number 1963 edition
Part I: General		
1	Interpretation, definitions, etc.	–
2	Contractor's obligations	1 ; 12(1) (2)
3	Contract sum – additions or deductions – adjustment – interim certificates	–
4	Architect's/supervising officer's instructions	2
5	Contract documents – other documents – issue of certificates	3
6	Statutory obligations, notices, fees and charges	4
7	Levels and setting out of the Works	5
8	Materials, goods and workmanship to conform to description, testing and inspection	6
9	Royalties and patent rights	7
10	Person-in-charge	8

Clause number 1980 edition	Clause heading	Clause number 1963 edition
11	Access for architect/supervising officer to the works	9
12	Clerk of works	10
13	Variations and provisional sums	11
14	Contract sum	13; 12 (part)
15	Value added tax – supplemental provisions	13A
16	Materials and goods unfixed or off-site	14
17	Practical completion and defects liability	15
18	Partial possession by employer	16
19	Assignment and sub-contracts	17
19A	Fair wages (local authority edition only)	17A
20	Injury to persons and property and employer's indemnity	18
21	Insurance against injury to persons and property	19
22A	Insurance of the works against clause 22 perils	20A
22B	Clause 22 perils – sole risk of employer	20B
22C	Clause 22 perils – existing structures – sole risk of employer	20C
23	Date of possession, completion and postponement	21
24	Damages for non-completion	22
25	Extension of time	23
26	Loss and expense caused by matters materially affecting regular progress of the works	24
27	Determination by employer	25
28	Determination by contractor	26
29	Works by employer or persons employed or engaged by employer	29
30	Certificates and payments	30
31	Finance (No 2) Act 1975 – statutory tax deduction scheme	30B
32	Outbreak of hostilities	32
33	War damage	33
34	Antiquities	34
	Arbitration (1980 edition – Article 5)	35

Part II: Nominated subcontractors and nominated suppliers

35	Nominated subcontractors	27
36	Nominated suppliers	28

Clause number 1980 edition	Clause heading	Clause number 1963 edition
Part III: Fluctuations		
37	Choice of fluctuation provisions – entry in appendix	
38	Contribution, levy and tax fluctuations	31B,C,D,E
39	Labour and materials cost and tax fluctuations	31A,C,D,E
40	Use of price adjustment formulae	31F
Appendix		
VAT agreement		

APPENDIX E

HISTORICAL NOTE

In 1903 agreed conditions of contract were published under the sanction of the RIBA, with the agreement of the IOB and NFBTE. Six years later the first edition of what was described as the RIBA Form of Building Contract appeared under the authority of the RIBA and NFBTE. From this beginning, nationally acceptable and stable contract conditions for carrying out building work have evolved, which are generally regarded as fair and reasonable by the parties concerned.

At the time of the revision of The RIBA Form of Building Contract in 1931, the JCT was set up (*see* Appendix B), and subsequent editions under its authority followed. In 1937 a new form for use by local authorities was published, including a fair wages clause. In 1939 independent arbitration on all points was provided for, and in 1963 and 1980 major revisions were completed leading to new editions.

Participation in the work of agreeing acceptable contract conditions grew with the support of the RICS which joined the JCT in 1952, the Local Authority Associations in 1957 and the Specialist Federations in 1966/7.

The first standard form of subcontract appeared in 1936. It was not until 1980 that the JCT published its Nominated Subcontract Form, the nominated form up to that date having been published by the NFBTE and the specialist organisations FASS and CASEC who continue to publish the 'blue' non-nominated form.

APPENDIX F

THE NATIONAL JOINT CONSULTATIVE COMMITTEE FOR BUILDING

The NJCC was formed in 1955 when the RIBA, RICS and NFBTE appointed representatives to the new body.

This advisory body is concerned with practice and procedure in building contracts. Although it has no executive powers and its decisions and publications can only be regarded as recommendations, its influence is growing in the field of tendering procedure and the letting and conduct of contracts. It provides a medium for discussing matters of interest to its constituent members, although it has to be careful not to trespass on the preserves of its constituent bodies or other organisations such as the JCT.

The Banwell Report in 1964 envisaged a more widely representative body in this field but it was not until 1973 that engineering and subcontracting interests were represented through ACE and the (NFBTE) Federation of Building Subcontractors. So far, the subcontracting organisation, FASS, has been unable to accept the membership terms proposed, although CASEC became a member in 1977.

In turn, each of the constituent bodies puts forward a chairman who is elected annually at the AGM and there is a permanent Secretariat. There is a Main Committee meeting each quarter, an Officers' Committee and a Good Practice Panel. Specialist Committees or Groups have been formed. Conferences and seminars are held from time to time and an AGM at the beginning of each year.

The most significant document published by the NJCC is the Code of Procedure for Selective Tendering which first appeared in 1959; the 1969 edition was revised in 1972 in collaboration with the DOE for use throughout the United Kingdom. In 1977 the code was further revised and published as the Code of Procedure for Single Stage Selective Tendering. A two-stage code is in preparation.

Some Regional Joint Consultative Committees or Boards have a longer history. The Eastern Counties Board, for example, was set up over 45 years ago.

There is now a close link and movement of ideas between the NJCC and the RJCCs, the Chairman of the former visiting the regions each year and the Chairman of RJCCs attending meetings of the NJCC Main Committee.

Publications

'Clients Guide'
'Management of Building Contracts'

'Tendering Procedures for Industrialised Building Contracts'
'Code of Procedure for Single Stage Selective Tendering' (1977)

Obtainable from RIBA, RICS and NFBTE – see Appendix I

Procedure Notes

1. Variations on Building Contracts
2. Alterations to Standard Form of Building Contract (revised March 1978)
3. Additional Information for Tenderers
4. Placing of Contracts with a Substantial Engineering Content
5. Selective Tendering – *withdrawn March 1978*
6. NEDO Price Adjustment Formula – *withdrawn March 1978*
7. Standard Form of Building Contract – Clause 31E (revised March 1978)
8. NEDO Price Fluctuation Formula – Clause 31F – *withdrawn March 1978*
9. Tendering Period (revised March 1978)
10. Advice to Tenderers : Use of Postal Services (revised March 1978)
11. Financial Control and Cash Flow
12. Tendering for Building Works without Bills of Quantities

These notes are obtainable free on receipt of stamped addressed envelope from the NJCC 18, Mansfield Street, London WI.

APPENDIX G

Price adjustment formula – comparison of work categories series 1 and 2

34 category (series 1) Reference 1/	48 category (series 2) Reference 2/	PSA work group	SMM section for 48 categories *5th Edition*	*6th Edition*
—	1. Demolitions	A	C	C
1. Site preparation, excavation & disposal	2. Site preparation, excavation & disposal	A	D, H, S, T, X	D, E, H, R, S, W
2. Hardcore & imported filling	3. Hardcore & imported filling	A	D, T, X	D, S, W
3. Piling	4. General piling	A	E	E
	5. Steel sheet piling	A	E	E
4. Insitu concrete	6. Concrete	B	F, H, S, T, X, Y	D, E, F, H, R, S, W, X
5. Reinforcement	7. Reinforcement	B	E, F, G, H, X	E, F, G, H, W
6. Precast concrete	8. Structural precast & prestressed concrete units	B	F	F
	9. Non-structural precast concrete components	B	F, T, U, X	F, S, T, W
7. Formwork	10. Formwork	B	F, H, X	E, F, H, W
8. Concretor sundries	—			
9. Brickwork				
10. Blockwork	11. Brickwork & blockwork	C	G, H, R, S, T, U, X, Y	G, H, R, S, T, W, X
11. Bricklayer sundries				
12. Natural stone	12. Natural stone	D	J, K, U, Y	J, K, T, X
13. Asphalt work	13. Asphalt work	E	L, U	G, H, L, T
14. Slate & tile roofing	14. Slate & tile roofing	E	M	M
	15. Asbestos cement sheet roofing & cladding	E	M, N, U, S	M, N, T, R
15. Sheet roofing & flashings	16. Plastic coated steel sheet roofing & cladding	E	M, N, U	M, N, T
	17. Aluminium sheet roofing & cladding	E	M, N, U	M, N, T
16. Built up felt roofing	18. Built-up felt roofing	E	M, U	M, T
	19. Built-up felt roofing on metal decking	E	M, U	M, T

17. Carpentry	20. Carpentry, manufactured boards & softwood flooring	F	M, N, P, U	M, N, R, S
	21. Hardwood flooring	F	P, U	N, R, S, T
18. Flooring	22. Tile & sheet flooring	J	U	N, T
	23. Jointless flooring	J	U	T
19. Softwood joinery	24. Softwood joinery	F	P	N
20. Hardwood joinery	25. Hardwood joinery	F	P	N
21. Manufactured boards	—			
22. Ironmongery	26. Ironmongery	F	P	N
23. Steelwork	27. Steelwork	G	Q	P
	28. Steel windows & doors	H	R	Q
24. Builders general metalwork & purpose made items	29. Aluminium windows & doors	H	R	Q
	30. Miscellaneous metalwork	H	R, S, T, X	Q, R, S, W
	31. Cast-iron pipes & fittings	I	S, X	R, W
	32. Plastic pipes & fittings	I	S	R
25. Gutters, pipework & associated fittings	33. Copper tubes, fittings & cylinders	I	S	R
	34. Mild steel pipes, fittings & tanks	I	S	R
	35. Boilers, pumps & radiators	I	S	R
26. Fittings, equipment & appliances	36. Sanitary fittings	I	S	R
27. Insulation & sundries	37. Insulation	I	N, S	N, R
	38. Plastering (all types) to walls & ceilings	J	U	T
28. Insitu finishings, beds & backings	39. Beds & screeds (all types) to floors, roofs & pavings	J	U	T
29. Tile slab block & plain sheet finishings	40. Dry partitions & linings	J	U	T
	41. Tiling & terrazzo work	J	U	T
	42. Suspended ceilings (dry construction)	J	U	T
30. Glass, mirrors & patent glazing	43. Glass, mirrors, patent glazing	K	V	U
31. Decorations	44. Decorations	L	S, T, W	R, S, V
32. Drainage goods	45. Drainage pipework (other than cast iron)	M	X	W
33. Fencing & gates	46. Fencing, gates & screens	M	Y	X
34. Roads & paths	47. Bituminous surfacing to roads & paths	M	L, U	L, T
—	48. Soft landscaping	M	D	D

APPENDIX H

UNFAIR CONTRACT TERMS ACT 1977

This Act, while affording protection to the consumer, affects business transactions including building contract conditions. This note very briefly considers some of the complicated provisions in relation to building operations.

The Act came into effect on 1 February, 1978 and applies to contracts entered into on and after that date and to any loss or damage suffered thereafter.

It imposes limits on the extent to which parties to contracts can exclude or restrict liability for breach of contract, negligence (failure to exercise reasonable care or skill) or other breach of duty. This includes the terms of non-contractual notices. Secondary contracts cannot be used to evade liability.

It does not apply to purely private transactions between individuals but it applies to business contracts where one party uses 'written standard terms of business' and to a contract where a 'consumer' deals with someone in his business capacity. 'Business' includes the professions and the activities of Government Departments and public authorities.

The numerous building contract conditions reviewed in this book would appear to come within the definition of 'written standard terms of business'. This and the reasonableness of some provisions may one day be tested in the courts, particularly on unilateral conditions, amendments to Standard terms, or restrictions of common law rights.

There must be a contractual liability and an attempt to exclude or restrict it. When this actually takes place the remedy is for the aggrieved party to go to Court to have the offending clause struck out. This action could only be taken when this situation arises not when the contract is entered into.

The restrictions at which the Act is aimed would include making the liability or its enforcement subject to onerous conditions (eg, unreasonable time limits), excluding or restricting remedies, or limiting the operation of the legal process. An arbitration agreement would not fall under this heading.

Exemption clauses on liability for breach, etc cannot therefore be enforced unless they satisfy the test of what was fair and reasonable at the time the contract was entered into or if there is damage through negligence, when the damage occurs. (The Misrepresentation Act 1967 is amended to reflect this test of reasonableness.) The onus of proof is on the claimant that the terms are, in fact, unreasonable.

Contracts of insurance are exempt and this exemption extends to land transactions, patents, company and partnership formation and dissolution and securities transactions. The extent of the exemptions is strictly limited so that in a contract to sell a house and land, the Act's exemption would apply to the land transaction but not the building contract.

A person cannot exclude or restrict his personal liability either through contract terms or by a non-contractual notice, for death or personal injury caused by negligence or breach of duty. This would extend to building operations.

In other instances the test of reasonableness is applied to attempts to restrict or exclude liability for negligence causing loss or damage.

With professional firms and small businesses in mind, where the liability is by reference to a specific sum of money, reasonableness will be judged by the resources available to meet the liability and the availability of insurance cover. A consumer cannot be required unreasonably to indemnify another person in respect of the liability of the other party for negligence or breach.

In relation to the sale of goods this Act extends the concept of the Supply of Goods (Implied Terms) Act which, as now amended, renders void in the case of consumers conditions of sale exempting implied obligations as to quality, fitness, etc and in other instances applies the test of reasonableness to their enforcement.

APPENDIX I

BUILDING CONTRACT FORMS AND OTHER PUBLICATIONS

Building contract forms and agreements issued by the Joint Contracts Tribunal (as at January 1980)

STANDARD FORM OF BUILDING CONTRACT – 1980 EDITION (referred to in Chapter 3)

Private Edition – with Quantities
Private Edition – without Quantities
Private Edition – with Approximate Quantities
Local Authorities Edition – with Quantities
Local Authorities Edition – without Quantities
Local Authorities Edition – with Approximate Quantities
Sectional Completion Supplement
Standard Form of Tender for Nominated Suppliers (Chapter 7)

STANDARD FORM OF NOMINATED SUBCONTRACT – 1980 EDITION (Chapter 6)

Standard Form of Nominated Subcontract Tender and Agreement	*Tender NSC/1*
Standard Form of Employer/Nominated Subcontractor Agreement	*Agreement NSC/2*
Standard Form for Nomination of a subcontractor where Tender NSC/1 has been used	*Nomination NSC/3*
Standard Form of Subcontract for subcontractors who have tendered on Tender NSC/1 and executed Agreement NSC/2 and been nominated by nomination NSC/3	*NSC/4*

Notes on the use of Tender NSC/1 and Agreement NSC/2 are published on the front of pads of NSC/1 (in sets of 3).

Agreement NSC/2 adapted for use where Tender NSC/1 has not been used	*Agreement NSC/2a*
Subcontract NSC/4 adapted for use where NSC/1, 2 and 3 have not been used	*Subcontract NSC/4a*

Explanatory Memorandum on Standard Form and Subcontract Form (this is published in a convenient package with the Standard Form).

FORMULA RULES (Chapter 4)
There is now published in one document:

(i) Standard Form of Building Contract Formula Rules – Work Category Indices Series 2 – 4 April 1977 (revised 1980)
(ii) Nominated Subcontract Formula Rules – Series 2 – 4 April 1977 with revisions suitable for use with 1980 Contract NSC/4 and 4a
(iii) Formula Rules for use with the Standard Form with Contractor's Design (1980)

OTHER JCT FORMS (Chapter 7)
Design and Build:

– Standard Form with Contractor's Design (1980)
– Clauses modifying Standard Form (1980 Local Authorities Edition with Quantities) for use with quantities and contractor's proposals

Fixed Fee Form of Prime Cost Contract (under revision)
Agreement for Minor Building Works (1980 edition)
Agreement for Renovation Grant Work:
(a) where architect appointed
(b) where no architect appointed
Practice Notes on Standard Form of Contract (pre-1980 edition) (Appendix C)
Form of Tender for Nominated Subcontractor (pre–1980 edition) (Chapter 5)

Note: JCT forms are obtainable from:
Royal Institute of British Architects,
66 Portland Place, London WI
Royal Institution of Chartered Surveyors,
12 Great George Street, London SWI
National Federation of Building Trades Employers,
82 New Cavendish Street, London WI

Publications available from NFBTE

Standard Form of Subcontract (nominated) April 1978 NFBTE/FASS/CASEC (Chapter 5)
Standard Form of Subcontract (non-nominated) July 1978 (revised 1980) (Chapter 5)
Design and Build Form (NFBTE) (January 1974) (Chapter 8)

Labour-only Form of Subcontract (1976)
(Chapter 8)
Form of Tender for Nominated Suppliers (also from RIBA and RICS)
(Chapter 8) (Clause 28 pre-1980 Standard Form)
Model conditions of estimate for small works (1979)
(Chapter 8)
Model Form of Enquiry for subcontract works 1976 (non-nominated)
(Chapter 5)
Agreement for the Construction of New Streets
(Chapter 8)
Definition of Prime Cost of Daywork carried out under a Building Contract (1975 – RICS/NFBTE; revised 1979)
(Chapter 8)
Definition of Prime Cost of Building Works of a Jobbing or Maintenance Character (this 1966 RICS/NFBTE publication is under review for early publication)
Standard Method of Measurement of Building Works (also from RICS) (6th edition July 1978)
(Chapter 2)
Code for the Measurement of Building Work in Small Dwellings (3rd edition 1979) (also from RICS)
(Chapter 2)

OTHER PUBLICATIONS
New Work Form (where no architect employed)
(Chapter 8)
(obtainable from Eastern Builders Federation, 95 Tenison Road, Cambridge)
Employer/Nominated Subcontractor Agreement 1973 (RIBA)
(Chapter 5)
Employer/Nominated Supplier Warranty Form (RIBA)
(Chapter 8)
National Building Specification
(Chapter 2)
(obtainable from RIBA, 66 Portland Place, London WI)
Schedule of Plant Charges (2nd edition July 1975) (under review)
(Chapter 8)
(obtainable from RICS, 12 Great George Street, London SWI)

NJCC PUBLICATIONS are listed in Appendix F and are available from 18 Mansfield Street, London WI.

PRICE ADJUSTMENT FORMULAE
In addition to publications listed under JCT above, the following are also available:

Guide to application and procedure (series 2) (HMSO)
Description of the indices (series 2) (HMSO)

NFBTE General Guide to the NEDO Formula Method (1975)
(Chapter 4)
The Sensitivity of the Building Price Adjustment Formula (NFBTE)
Formula Rules – Series 2 – 4 April 1977
(1) Main contract
(2) Subcontracts (nominated 'green' and non-nominated 'blue')
Monthly Bulletin (Construction Indices) Building Works (HMSO)
(Chapter 4)

Prices are not quoted since these are subject to frequent change.

Note: Civil engineering forms, which do not come within the scope of this book, are obtainable from the Federation of Civil Engineering Contractors, Romney House, Tufton Street, London SWI. They are:

(a) ICE General Conditions of Contract
(b) Form of Subcontract designed for use in conjunction with the ICE General Conditions of Contract.

INDEX

To facilitate reference to various points this index primarily deals with items under the particular form of contract (eg, standard form of building contract) with additional and separate limited references to main subject matter (eg, arbitration).